# Nanobioceramics for Healthcare Applications

# Nanobioceramics for Healthcare Applications


Editors

**Eng San Thian**
NUS, Singapore

**Jie Huang**
University College London, UK

**Mamoru Aizawa**
Meiji University, Japan



World Scientific

NEW JERSEY • LONDON • SINGAPORE • BEIJING • SHANGHAI • HONG KONG • TAIPEI • CHENNAI • TOKYO

*Published by*

World Scientific Publishing Europe Ltd.
57 Shelton Street, Covent Garden, London WC2H 9HE
*Head office:* 5 Toh Tuck Link, Singapore 596224
*USA office:* 27 Warren Street, Suite 401-402, Hackensack, NJ 07601

**Library of Congress Cataloging-in-Publication Data**
Names: Thian, Eng San, editor. | Huang, Jie (Biomedical engineer), editor. |
Aizawa, Mamoru, 1968– editor.
Title: Nanobioceramics for healthcare applications / [edited by] Eng San Thian (NUS, Singapore), Jie Huang (University College London, UK) & Mamoru Aizawa (Meiji University, Japan).
Description: New Jersey : World Scientific, 2016. |
Identifiers: LCCN 2016023264 | ISBN 9781786341334 (hc : alk. paper)
Subjects: | MESH: Ceramics--therapeutic use | Nanostructures--utilization |
Orthopedic Fixation Devices
Classification: LCC R857.M3 | NLM QT 37.5.C4 | DDC 610.28--dc23
LC record available at https://lccn.loc.gov/2016023264

**British Library Cataloguing-in-Publication Data**
A catalogue record for this book is available from the British Library.



Desk Editors: Anthony Alexander/Mary Simpson

Typeset by Stallion Press
Email: enquiries@stallionpress.com

Printed in Singapore

# Preface

Maintaining a good quality of life in an ageing and active population is one of the great challenges in the 21st century. Research into bioceramics for healthcare applications has started as early as 1970s; and over the past 40 years, it has greatly expanded. This book is specifically written to fill the gap in the literature by providing a comprehensive discussion on nanobioceramics for healthcare applications, a field that is advancing at a very rapid pace.

This book contains nine chapters, all of which are written by experts in the field of bioceramics from the United States, United Kingdom, Japan, Australia, Singapore, and China. Each chapter presents an in-depth coverage of various nanobioceramics via the most up-to-date findings. Chapter 1 gives a broad overview of nanobioceramics; Chapter 2 introduces silica-based nanoceramics; Chapter 3 focuses on hydroxyapatite nanoceramics; Chapter 4 talks about tricalcium phosphate-based nanoceramics; Chapter 5 deals with calcium phosphate nanocoatings; Chapter 6 covers bone nanocements; Chapter 7 mentions about bioglass-based nanoceramics; Chapter 8 discusses about glass-ceramics; and Chapter 9 describes composite-based nanoceramics. For those who wish to know more can refer to the detailed list of references cited at the end of each chapter.

I would like to thank all the authors for their great contribution to this informative book. Also special thanks to the reviewers who provided invaluable critique to each chapter. Finally, I would like to acknowledge the staff at World Scientific Publishing, especially Merlin Fox, Catharina Weijman, Mary Simpson, and Anthony Alexander, for their continuous assistance that made the publication of this book a reality.

Eng San Thian
Jie Huang
Mamoru Aizawa

# About the Editors

Dr. Eng San Thian is an Associate Professor of Biomedical Materials in the Department of Mechanical Engineering, and a Principal Investigator in the Centre for Biomedical Materials Applications & Technology at National University of Singapore.

Dr. Jie Huang is a Reader of Biomaterials and heads the Biomedical Materials Laboratory at the Department of Mechanical Engineering, University College London.

Dr. Mamoru Aizawa is a Full Professor in the Department of Applied Chemistry, School of Science and Technology, and leads a research group in the Laboratory of Biomaterials at Meiji University.

# Contents

*Preface* v

*About the Editors* vii

Chapter 1 Overview of Nanobioceramics 1
*Poon Nian Lim, Jie Huang, Mamoru Aizawa and Eng San Thian*

Chapter 2 Silica-based Nanoceramics 23
*Akiko Obata and Toshihiro Kasuga*

Chapter 3 Bioceramics for Orthopaedic Device Applications: Hydroxyapatite 49
*Qiaomu Tian, Laura Rivera-Castaneda, Zachary S. Dunn and Huinan Liu*

Chapter 4 Tricalcium Phosphate-based Nanoceramics 79
*Kazuaki Hashimoto*

Chapter 5 Calcium Phosphate Nanocoatings: Production, Physical and Biological Properties, and Biomedical Applications 105
*Besim Ben-Nissan and Andy H. Choi*

Chapter 6 Bone Cements Utilised for the Reconstruction of Hard Tissue: Basic Understanding and Recent Topics 151
*Kanji Tsuru, Yuki Sugiura and Kunio Ishikawa*

Chapter 7 Strontium-Containing Mesoporous Bioactive Glass for Regeneration of Osteoporotic Bone and Periodontal Tissue 187
*Chengtie Wu and Jiang Chang*

Chapter 8 Bioactive Glass-Ceramics 213
*Taishi Yokoi, Toshiki Miyazaki, Masakazu Kawashita and Chikara Ohtsuki*

Chapter 9 Nanocomposites for Bone Repair 239
*Kyung-Ah Kwon, Jennifer H. Shepherd, David V. Shepherd, Ismael Moreno-Gomez and Serena M. Best*

*Index* 299

# Chapter 1

# Overview of Nanobioceramics

**Poon Nian Lim*,¶, Jie Huang†,||**
**Mamoru Aizawa‡,** and Eng San Thian*,††**

*Department of Mechanical Engineering, National University of Singapore, 9 Engineering Drive 1, Singapore 117576, Singapore*

†*Department of Mechanical Engineering, University of London, London WC1E 7JE, UK*

‡*Department of Applied Chemistry, School of Science and Technology, Meiji University, 1-1-1 Higashimita, Tama-ku, Kawasaki, Kanagawa 214-8571, Japan*

¶*poonnian@gmail.com*, ||*jie.huang@ucl.ac.uk*,
***mamorua@isc.meiji.ac.jp*, ††*mpetes@nus.edu.sg*

**Abstract**

Bone is the second most commonly transplanted organ in the human body. The concept of repairing bone injuries has evolved greatly over the last five decades to the third generation of biomaterials. One of the emerging areas of research in the third generation of biomaterials is the integration of nanotechnology into bioceramics. Due to its close chemical similarity to the human bone mineral, nanobioceramic has emerged as a new generation of biomaterials, which aims to create a micro-environment that induces enhanced cell responses leading to osteogenesis. This chapter will thus present an overview of the general characteristics of nanobioceramics. The future prospects and challenges of nanobioceramics will also be covered.

## 1.1 Biomaterials for Bone Repair

Bone is a dynamic and complex tissue that provides structural support and withstands load-bearing for the body. Moreover, it also serves as a reservoir for minerals, supports muscular contraction for mechanical movement, and protects the internal organs.[1] Through undergoing remodelling, bone has the innate capacity to regenerate and repair itself upon damage. However, when huge amount of bone is lost in circumstances such as trauma, diseases, or old age, the function of bone is adversely affected. Under such circumstances, bone function can only be restored by surgical reconstruction. Bone repair is a procedure of replacing the defect part with material from either the patient's own body (autografting) or that of a donor (allografting).[2,3] Autografts contain living bone cells and growth factors that offer osteogenesis and stimulate induction, hence providing bone ingrowth. Even though autograft is considered the "gold standard", only a limited amount of bone can be harvested for use in bone repair. In relation to skeletal reconstructive surgery, an alternative has been the use of donated allogenic bone. However, processing (freezing or demineralisation) of the allografts adversely affects the structural integrity of the material and its osteoinductive and osteogenic activity. Hence, its implementation for orthopaedic applications has been hindered by the incomplete remodelling and limited availability. Furthermore, both autografting and allografting involve multiple surgeries and are often associated with the risk of a series of complications such as infection, hematoma, cosmetic disadvantages, post-operative pain, and chronic pain at the donor site.[4,5] Thus, autografts and allografts do not always provide the best solution for bone repair. Therefore, all these factors greatly drive the market for synthetic bone alternatives.

The concept of placing implants into the human body has evolved greatly over the last five decades. It began in the 1960s with the first generation of biomaterials designed to achieve matching mechanical properties of the replaced tissues, with a minimal toxic response in the host. At that time, bioinert materials are deemed biocompatible and have been used in the following two areas: (1) permanent prostheses such as hip prosthesis and dental implants and (2) fixation tools for bone fractures such as plates, pins, screws, and rods. Bioinert implants such as stainless steels, cobalt–chromium alloys, titanium alloys, alumina, and zirconia are able to provide mechanical support in load-bearing applications.

However, the body regards these bioinert materials as foreign, triggering fibroblasts to form a fibrous capsule around the implant. The formation of a protective fibrous layer isolated the implant from the host tissue, which created a separation between the implant and the host.[6] Thus, bioinert implants could not be integrated into the bone tissue. Furthermore, there were also concerns regarding undesirable tissue responses caused by the released ions and small debris particles from the metallic and ceramics implants, respectively.[7] In addition to the issue of undesirable responses, the high mechanical stiffness of metallic implants often resulted in stress shielding and thus bone resorption due to the mismatch of elastic modulus between the implant and bone.[8] As a consequence, a second operation is often required to remove the failed bioinert implant. Hence, it soon becomes apparent that bioinert materials are not adequate in eliciting full tissue growth and regeneration.

With this in mind, material scientists begin to partner with physicians to develop the second generation of biomaterials. In the 1980s, this has progressed towards biomaterials that can elicit a controlled action and reaction in the physiological environment, conferring a bioactive or biodegradable behaviour. The interface problem can be resolved with bioactive or biodegradable materials. Bone can directly form an interfacial "bond" with the surface of bioactive materials with no mediation of a fibrous connective tissue interface and further undergo mineralisation. Biodegradable materials exhibit clinically relevant, controlled chemical breakdown and resorption, which can be ultimately replaced by regenerating tissues. During the last 40 years, bioceramics have received significant attention for biomedical applications due to their ability to interact with human biological system. With these favourable biological responses, a wide array of ceramic materials have been used. This ranges from chemically pure oxides to various calcium phosphate-based ceramics,[9] bioactive glasses,[10] glass-ceramics,[11] and the chemically and microstructurally complex ceramic-matrix composites.[12] These bioceramics can be fabricated in a variety of forms to serve various functions in the body repair.[13]

Calcium phosphate-based ceramics can be fabricated into porous implants, granules, powders, and coatings on metallic implants. For instance, hydroxyapatite (HA) coatings provided osteoconduction in which bones grew onto the coatings and formed mechanically strong interface.[13] Bioactive glasses and glass-ceramics were used as middle-ear prostheses, oral implants, and vertebrae replacement for spinal tumours.[14] Bioactive composite consisting of HA particles in a polyethylene matrix has become important in the repair and replacement of bones in the middle ear.[15]

Despite the clinical success of bioinert, bioactive, and biodegradable implants in meeting the medical needs of the rapidly aging population, 30–50% of the prostheses failed within 10–25 years, leading to revision surgeries.[16] Unlike living tissues, synthetic materials cannot respond to the changing physiological loads or biochemical stimuli. Hence, this limits the lifetime of the synthetic implants. Fifty years of research in the advancement of first- and second-generation biomaterials are limited in the repair and restoration of the body and this signals the need for a shift of the current medical paradigm towards a more biological-based approach for the regeneration of tissues.

This shift occurs at the beginning of 2000s, which brings us to the third generation of biomaterials, the cell- and gene-activating biomaterials that are designed to stimulate specific cellular responses at the molecular level. Third-generation materials are not meant to substitute plainly the materials from previous generations. Instead, they improve the initial physiological cues and bioactivity existing in the second-generation biomaterials, to allow the biomaterials and surrounding environment to adapt, signal, and stimulate specific cellular activity and behaviour in order to regenerate new tissues. The aim to regenerate rather than repair leads to the exploration of tissue engineering and regenerative medicine by using natural signalling pathways and components such as stem cells, growth factors, and peptide sequences, in combination with the synthetic scaffold. Artificial tissues are being fabricated by placing cells within scaffold materials so as to guide cell proliferation and differentiation. Hence, this opens up new possibilities of treatments and applications.

Along this new wave of thinking, one of the emerging areas of research is the integration of nanotechnology into the field of biomaterial science. Nanotechnology has opened up a new era of technology, particularly the development of nanostructuring and nanoscale engineering of the biomaterials. The intriguing properties of nanobiomaterials are explored in several ways. Metallic nanoparticles have found significant applications in a wide spectrum of biomedical utilities such as imaging, sensing, drug delivery, and gene targeting.[17–20] Carbon-based nanomaterials including fullerenes, nanotubes, nanodiamonds, and graphene show great promise as a carrier for pharmaceutical agents, components in nanoformulations for the delivery of therapeutic molecules, biosensor, imaging, drug delivery, bacterial inhibition, and photothermal therapy.[21–23] Polymeric nanoparticles have been used as drug carriers,[24,25] and nanobioceramics have been used extensively in orthopaedic and cranio-maxillofacial applications.[26] Indeed, nanotechnology has evolved rapidly from the discovery phase to the application phase; its potential impact on the healthcare is immense and has pervaded many aspects of the medical field (Table 1.1).[27]

Driven by the increasing number of patients afflicted with traumatic and non-traumatic conditions, the demand for the reconstructive and regenerative surgeries in orthopaedic, cranio-maxillofacial, and dental applications is growing. Due to its close chemical similarity to the human bone minerals, bioceramics play a crucial role in the reconstructive and regenerative fields for the

Table 1.1 Medical application of nanotechnology. Reprinted with permission from Ref. 27.

| Application | Medical use |
|---|---|
| Nanovectors | Targeted delivery of genes to selected disease sites |
| Nano-electrotransducers | Electrostimulation therapies |
| Control of cell movement | Cell therapy for regeneration and tissue engineering, islet cell transplantation, anti-metastasis tumour therapy |
| Magnetic drug delivery | Controlled release targeted drug delivery |
| MRI contrast agents | Imaging of tumours and molecular imaging |
| Nanohyperthermic ablation | *In situ* destruction of tumours as an alternative as a complement to surgical excision |
| Antibacterial | Antisepsis agents, therapy for antibiotic-resistant bacteria, infection-resistant bioimplants, stents, and prostheses |
| Electrochemotherapy | Low-voltage non-contact electrochemotherapy of tumours avoid the use of external electrodes |
| Nanobiosensors | *In vitro* and *in vivo* diagnostic |
| Nanoscaffolds | Regenerative medicine and tissue engineering |

second generation of biomaterials. With the advent of nanotechnology, nanobioceramics emerge as a new generation of biomaterials, which aim to create a microenvironment that induces a cell response leading to osteogenesis. This chapter will present an overview of the characteristics of nanobioceramics, followed by the lesson learnt from nanobioceramics. Lastly, the future prospects and challenges of nanobioceramics will be covered.

## 1.2 Characteristics of Nanobioceramics

In general, nanomaterials are defined as materials that possess constituents of less than 100 nm in at least one direction.[28] Nanobioceramics including $\alpha$-alumina, zirconia, calcium phosphates, and bioglasses encompass many different forms, which range from bulk materials comprising basic structural units in the range of 1–100 nm (nanostructured), crystalline solid with grain sizes in the range of 1–100 nm (nanocrystals), individual layer or multilayer surface coatings with thicknesses in the range of 1–100 nm (nanocoatings), extremely fine powders with average particles sizes in the range of 1–100 nm (nanopowders), to fibres and rods with diameters in the range of 1–100 nm (nanofibres and nanorods).

Similar to bioceramics, nanobioceramics are highly biocompatible, stable in the physiological environment, corrosion-resistant, and have low thermal and electrical conductivities. Due to the unique chemical structure, nanobioceramics are also susceptible to surface and chemical modifications. Thus, nanobioceramics retain most of the inherent properties of bioceramics. However, at the nanoscale, some of the properties of nanobioceramics become very different from their conventional counterparts, and this leads nanobioceramics to exhibit a number of distinctive behaviours relative to bulk materials, which are predominantly reflected in the physical, surface, and mechanical properties.

The size of nanobioceramics is below one-fourth the wavelength of light; hence they do not scatter light, which gives rise to transparency. The transparent property resulting from reduced microstructure of nanobioceramics can be exploited to the aesthetic benefits in the dental applications. Nanobioceramics contain greater amount of grain boundaries than conventional forms, providing unique surface properties such as topography, energy, roughness, and wettability that potentially exert a significant influence on the cell behaviour.[29–36]

When compared to a surface with microscale feature, one of the most obvious advantages of a nanostructured surface was its remarkably higher specific surface area. Nanostructured surface had an increased number of atoms and crystal grains (Figure 1.1),[29] which would become more active in terms of dissolution and recrystallisation processes. This allowed nanobioceramics to serve as a potential drug-delivery system for a variety of remedies such as antibiotics, anti-tumour and anti-inflammatory drugs, proteins, growth factors, and genes.[37–39] In addition, nanobioceramics

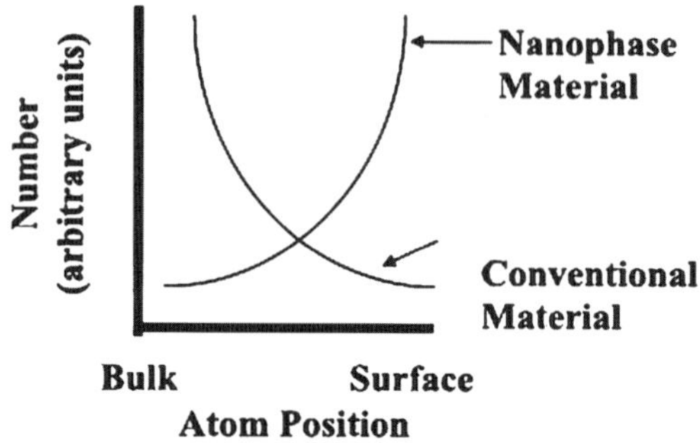


Figure 1.1 Number of atoms at the surface of nanophase and conventional materials. With kind permission from Ref. 29.

possessed a higher surface area-to-volume ratio than conventional microscale bioceramics, which would provide more available sites for protein adsorption, drug loading, and surface modification. Moreover, the high surface area in nanobioceramics altered the corresponding surface energy for protein adsorption, which would in turn facilitated the amount of cellular interactions.

On the flip side, the large surface area in nanobioceramics brings about strong driving force for the nanoparticles to form agglomerates as the material try to minimise the total interfacial energy of the system. This poses difficulty in handling nanoparticles as well as resulting in inhomogeneous properties when nanobioceramics are added to another material. To disperse the nanoparticles, appropriate controls including addition of surfactants and electrostatic repulsion agents during the synthesis stage and processing steps are required.

The surfaces of the micrometre-size particles are smooth at nanoscale. On the contrary, when the grain size and diameter of surface pore are reduced to nanoscale, they result in high surface roughness. Surface roughness was found to increase by 35–50% on nanophase alumina and HA as compared to their conventional grain size.[32,35,36] Due to greater surface roughness and greater numbers of grain boundaries on the surfaces of nanobioceramics, surface wettability is also enhanced. This was demonstrated when there was a three times reduction in the aqueous contact angles when alumina grain size was reduced from 167 to 24 nm.[35] The first step of cell adhesion involves protein adsorption onto the material surface.[32,34–36] Thus, the characteristics of nanometre topography and enhanced surface wettability in nanobioceramics will aid in the availability of the proteins to promote cell adhesion, protein adsorption, and protein unfolding.

Brittleness is a major drawback in bioceramics for repairing large bone defects. However, based on Hall–Petch relation, when grains are refined to nanoscale regime, it gives rise to notable increment in the hardness and yield strength of nanobioceramics. However, when grain size is below ~20 nm, nanocrystalline grains contain huge amount of triple grain boundary junctions, which reduces the strength. The reduction in strength can be explained by several mechanisms which encourage deformation in nanoceramics by the enhanced diffusion along grain boundary, grain boundary sliding, and microcracking.[40,41] Generally, the size and distribution of the critical flaws in ceramics affect the fracture strength. By maintaining a nanocrystalline microstructure for bioceramics, the homogeneity of the microstructure is improved, flaw sizes are reduced, thermal residual stresses arises from anisotropic crystal structure, and the coefficient of thermal expansion mismatch among microstructural phase assemblage are also lowered. Collectively, all these effects result in a significant higher stress for crack propagation, which give rise to a higher fracture strength and fracture toughness at the finest sizes of nanobioceramics.

This was demonstrated clearly in the increment of the strength of nano-HA when the defect size was reduced (Figure 1.2).[29] Despite becoming mechanically robust, it has been experimentally observed that the fracture toughness of a single-phase nanobioceramics was still inadequate for orthopaedic applications.[29] Nevertheless, this drawback can be overcome by incorporating nanobioceramics as reinforcement with a tough material. With nanobioceramics reinforced in the nanocomposite, the fracture toughness of the nanocomposite was improved to a moderate extent that could possibly support and protect the bone tissue construct during implantation procedure and within the

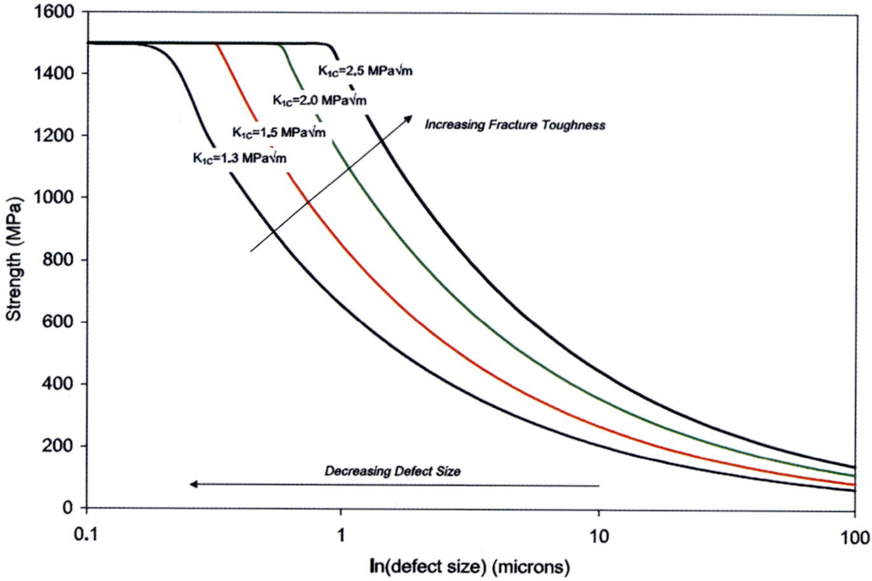


Figure 1.2 Effect of defect size on the strength of nano-HA. With kind permission from Ref. 29.

body.[42–44] Hence, these enabling properties make nanobioceramics an attractive platform for bone repair.

Taken together, the synergistic effects of improved surface and mechanical properties render nanobioceramics as a promising candidate to improve a series of requirements of a bone implant. In fact, nanobioceramics based on bioactive materials are deemed superior for dental and orthopaedic implants, owing to the better possibility of controlling the compositions, grain size distributions, surface and mechanical properties, to more closely match those of the physiological bone.

## 1.3 Lesson Learnt from Nanobioceramics

The concept of manipulating the properties of biomaterials has changed dramatically during the past five decades. In the early days, biomaterials are used based on the inherent properties. Today, nanotechnology offers tremendous opportunities to alter the properties of biomaterials to adjust the interaction with biological entities. Over 15 years, nanobioceramics play an increasingly crucial role in bone science. The unique characteristics of nanobioceramics have addressed several limitations of conventional bioceramics. One important lesson, which we can learn from nanobioceramics, is that it is possible to direct and control the cell behaviours by creating similarities between biomaterials and biological entities.

This was first noticed when an interfacial "bond" was formed directly on the surface of HA, and cells adhered onto HA further underwent mineralisation *in vivo*.[45] Following on from the promising results, an improvement in the rate of bone formation was demonstrated when cells were cultured with silicon-substituted HA and bioglass.[46,47] All these results demonstrated that by mimicking the composition of biomaterial closer to bone, it promoted favourable biological responses.

Thus, bioceramics become the forefront of biomaterials as they possess the closest composition of the natural bone. However, natural bone is highly complex and is characterised by a hierarchical structure. Imitating the composition factor alone is not sufficient to promote excellent bone regeneration. Living systems are governed by molecular behaviour at nanometre scales.[48,49] Cell adhesion is dependent on the selective adsorption of specific extracellular matrix proteins (e.g., fibronectin, vitronectin, bone sialoprotein, type I collagen, and laminin) onto the region of the biomaterial. Specific extracellular matrix proteins interact with cells via specific ligand–receptor interaction, followed by triggering a cascade of signalling pathways as a result of the formation of focal complex by the aggregation of intracellular biomolecules near the focal adhesion site. These signalling pathways thus regulate the cell behaviour. All these focal adhesion and integrin-mediated cell attachment function at the nanometre range. Having understood the considerable difference between each step of the cell interaction, this necessitates the development of nanobioceramics that are optimally engineered to mirror the innate physiological characteristics of bone so as to re-create a truly biomimetic microenvironment.

Several findings consistently testified the excellent ability of nanobioceramics to mimic the constituent components of natural bone, thereby promoting new bone formation.[31–36,50,51] The unique surface properties and energetics of nanobioceramics gave rise to nanofeatures including larger surface areas, higher surface roughness, and better wettability, which enhanced the interaction with proteins, particularly, fibronectin and vitronectin, which were known to influence the cell function of osteoblasts.[34] Studies reported that by decreasing the grain size below 100 nm, a 10% increase in the adsorption of vitronectin on alumina could be achieved. Furthermore, there was also an increased unfolding of the fibronectin and vitronectin on the nanophase ceramics as compared to conventional grain size ones.[32] Unfolding of these proteins promoted the availability of specific cell adhesive epitopes (such as the amino acid sequence Arg–Gly–Asp or RGD and those that recognised Lys–Arg–Ser–Arg or KRSR), which led to an increase in bone cell adhesion and function. These favourable interactions of fibronectin and vitronectin in turn promoted adhesion, proliferation, differentiation, and calcium deposition of osteoblasts on nanobioceramics.[32,35,36] Studies reported a four and two times increment of calcium–mineral deposition when osteoblasts were cultured on nanosized alumina and HA rather than on conventional counterparts, respectively.[33] In another study, cells were aligned on the surface of an implant when topographical features have been created by electrohydrodynamic printing with nano-HA.[51]

In addition to osteoblasts, the effects of nanotopography by nanobioceramics were also demonstrated on diverse cell types including mesenchymal stem cells[52,53] and osteoclasts.[54] There were a two times increment in the synthesis of tartrate-resistant acid phosphates (TRAP) by osteoclasts and the formation of pits on nano-HA as compared to the larger grain size HA.[54] The increase in the number of grain boundaries at the surface of smaller grain size nanobioceramics increased the diffusion of TRAP, leading to the formation of more resorption pits. As a result, nanobioceramics brought about coordinated function in osteoblasts and osteoclasts, which gave rise to an enhancing effect in the remodelling of the

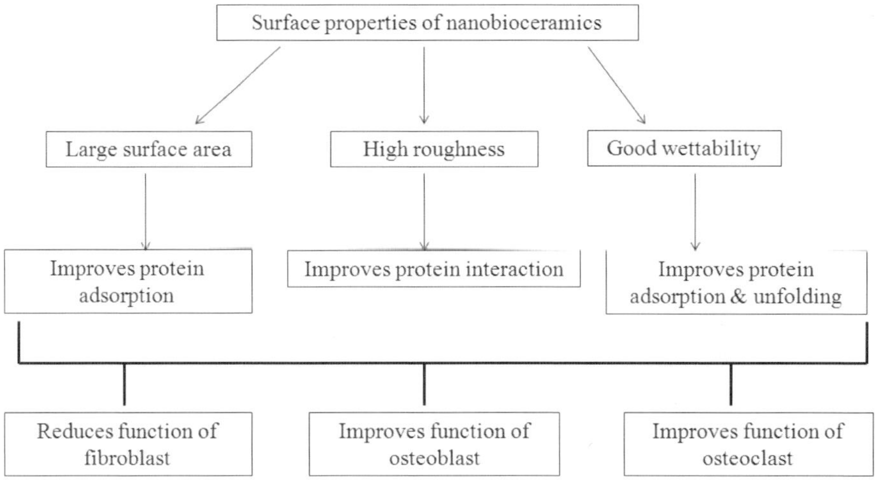


Figure 1.3 The influence of the nanofeatures in nanobioceramics for cell interaction.

newly formed bone juxtaposed to implants. Furthermore, it was shown that there was a decreasing function of the competitive cells such as fibroblasts (cells that contribute to fibrous encapsulation and callus formation events that might lead to implant loosening and failure) on nanophase ceramics as compared to conventional ones.[31,32] The influence of the nanofeatures, which nanobioceramics bring about for cell interaction, is summarised in Figure 1.3.

Zirconium oxide and titanium oxide are bioinert ceramics that possess good mechanical properties as a reinforcing phase to improve the fracture toughness and hardness of the material. Through nanostructuring of these materials, there was a possibility to transform these bioinert ceramics into bioactive ones.[33,55] Coatings of zirconium oxide and titanium oxide with nanosize grains (<50 nm) were demonstrated to induce the deposition of apatite in stimulated body fluid, whereas the same material with micro-size grain did not.[33,55] Furthermore, osteoblast adhesion and proliferation were enhanced by the nanostructured zirconium oxide coatings when compared to its micron-size counterpart.[56] The formation of nanostructured surface allowed zirconium oxide and titanium oxide to provide sufficient sites to induce the nucleation and growth of apatite. These results evidently demonstrated the importance of nanoscale feature on the physicochemical properties of biomaterial, thus influencing their biological responses.

HA is bioactive, but fails to exhibit osteogenesis efficiently. Since the initial interaction between cells and biomaterials were dependent on the cell-specific peptides, Babister *et al.*[57] utilised the high surface area characteristics in nanoparticles to functionalise RGD peptide onto HA nanoparticles. Peptide fragments could be conjugated onto HA nanoparticles and were more stable than incorporating growth factors and extracellular matrix proteins with HA. Cells interacted with the peptide/HA nanoparticles via electrostatic attraction were able to improve the osteogenic differentiation of human bone marrow cells in the non-osteogenic culture conditions. Furthermore, it was reported that when cells and peptide–HA nanoparticles were aggregated into micropellets, which gave rise to *in vivo* mineralisation and matrix production in 21 days of subcutaneous implantation in nude mice.[57]

In many cases, growth of cells discontinued in scaffolds at a thickness of only four to seven cell layers due to inadequate nutrient diffusion. Kretlow and Mikos[58] noted that scaffold materials with nanoporous structures containing pore sizes in the nanoscale regime (50–100 nm) were required in order to develop artificial tissues with appropriate sizes for clinical use. Therefore, it was essential to fabricate nanomaterials into three-dimensional scaffolds to achieve a better control of certain properties of biomaterials such as porosity and other mechanical properties that could influence cell behaviour. As such, many efforts have been dedicated to explore the influence of biomimetic nanofibrous scaffold using a composite material system.[59] To mimic the dimension, topography, and architecture of collagen fibrils in extracellular matrix, electrospinning technology was extensively researched.[60–62] To date, more than 100 different biodegradable polymers have been successfully electrospun and over 30 of them have been used for a variety of tissue-engineered applications.[61] Among the natural polymers, collagen, gelatine, elastin, silk fibroin, fibrinogen, hyaluronan, and chitosan have recently been fabricated into three-dimensional, nanofibrous scaffolds for orthopaedic applications.[61] The addition of nano-HA helped to improve the osteoconductivity and osteointegration of the biomimetic nanofibrous scaffolds, but the nanoparticles were prone to agglomeration and were difficult to disperse homogenously in an organic polymer solution.

Taken together, all these results demonstrate that nanobioceramics can possibly support favourable responses with biological entities to foster cell function and tissue growth. Therefore, it is becoming increasingly evident that the interaction between cells and microenvironment created by nanobioceramics can re-organise cytoskeleton and induce specific cell signalling. Ultimately, the development of a bone prosthesis using nanobioceramics will enable one to consciously incorporate properties to influence the physiological response at each stage of the bone healing process and as such, desired outcome can be obtained with greater consistency.

## 1.4 Future Prospect and Challenges of Nanobioceramics

On the whole, nanotechnology has much more to offer than just simple miniaturisation and building the molecular structures from the atomic scale. Over the last 20 years, numerous discoveries and applications of nanotechnology have revolutionised multiple disciplines of science, technology, and medicine. In particular, the application of nanotechnology to biomaterial science has become a new frontier for bone repair, offering promising scope to improve orthopaedic treatment, care, and prevention of bone diseases. Currently, great efforts are underway to optimise the structure and properties of nanobioceramics, to emulate the micro- and nanostructural elements and compositions of bone to obtain desirable characteristics. All these works have led research to reach the stage with early clinical trial. In the near future, with greater understanding of cell interactions, substantial work will be focused on the dispersion of the agglomerated nanobioceramics in order to exhibit its unique and useful properties for desired mechanical or biological effects, and improving angiogenesis within the nanobiomaterial system. Hence, it is envisaged that nanobioceramics will enable the development of new systems that mimic the complex hierarchical structure of the native tissue to a greater extent.

Although current investigation seems to support the impact of nanobioceramics in the domain of bone science, significant advancements are necessary to realise their full potential in clinical use. The following challenges and critical issues are required to be addressed in order for nanobioceramics to move on the next developmental phase.

One of the major challenges is the inconsistency of the processing technologies of nanostructured materials, which hinder the scale-up of nanofabrication processes. Nanofabrication processes incorporate a range of advanced processes, tools and materials which rely on top-down and bottom-up nanoscale processes. Manufacturing system issues in the product design, reliability and quality, process design and control, shop floor operations and supply chain management have yet to be optimised, which are the key barriers in advancing nanobioceramics from the laboratory into high volume production.

Another challenge is nanotoxicology, which is currently a relatively new field. Despite understanding the cellular and molecular basis that governs the interaction between nanostructured materials and cells, there are still huge gaps in our knowledge about the nature of the interactions of nanoparticles with the human system, in particular, the safety aspects of nanoparticles remains unaddressed. Recently, a report has indicated possible negative effects of carbon nanotubes on cells and their potential to provoke oxidative stress, inflammation, genetic damage and long-term pathological effects.[63] This may pose a threat to the human health as there may be circumstances when the interaction of nanobioceramics with biomolecules *in vivo* or nanobioceramics in their aggregation states may interact in an unwanted manner, triggering inflammatory responses. Furthermore, the concept of dosimetry, dose metrics exposure assessment, hazard identification and risk characterisation of nanobioceramics are still far from being in a conclusive state. Therefore, the influence of nanobioceramics on human health is still not well understood. Thus, substantial investigations are definitely needed to evaluate the stability of nanobioceramics in a variety of test systems so as to fully elucidate its use.

Indeed, the integration of biomaterial science and nanotechnology has clearly had an impact on bone repair. The research and development of nanobioceramics should continue, and in the near future, it will definitely bring about significant changes in the way bone repair is being approached. Hence, nanobioceramics open a wide range of opportunities and benefits to both the surgeons and the patients. If the raised issues can be overcome, nanobioceramics will revolutionise the treatment of bone repair. Overall, current trends in nanobioceramics foreshadow a bright future in the domain of bone science.

## References

1. Hing KA. Bone repair in the twenty-first century: Biology, chemistry or engineering? *Philos Trans R Soc Lond B Biol Sci* 2004;362:2821–2850.
2. Meeder PJ, Eggers C. The history of autogenous bone grafting. *Injury* 1994;25:SA2–SA4.
3. Czitrom AA, Gross KE. *Allografts in Orthopaedic Practice.* Baltimore: Williams & Wilkins; 1992; p. 7.
4. Togawa D, Bauer TW, Lieberman IH, Sakai H. Lumbar intervertebral body fusion cages: Histological evaluation of clinically failed cages retrieved from humans. *J Bone Joint Surg* 2004;86:70–79.
5. Barriga A, Díaz-de-Rada P, Barroso JL, Alfonso M, Lamata M, Hernáez S, Beguiristáin JL, San-Julián M, Villas C. Frozen cancellous bone allografts: Positive cultures of implanted grafts in posterior fusions of the spine. *Eur Spine J* 2004;13:152–156.
6. Heimke G. *Osseo-integrated Implants.* Boca Raton, FL: CRC Press; 1990; p. 31.
7. Black J. *Biological Performance of Materials.* New York: CRC Press; 1981; p. 233.
8. Daniels AU, Chang MKO, Andriano KP. Mechanical properties of biodegradable polymers and composites proposed for internal fixation of bone. *J Appl Biomater* 1990;1:57–78.

9. Jarcho M, Bolen CH, Thomas MB, Bobick J, Kay JF, Doremus RH. Hydroxylapatite synthesis and characterization in dense polycrystalline form. *J Mater Sci* 1976;11:2027–2035.
10. Hench LL. Bioceramics: From concept to clinic. *J Am Cer Soc* 1991; 74:1487–1510.
11. Kokubo T, Ito S, Huang ZT, Hayashi T, Sakka S, Kitsugi T, Yamamuro T. Ca,P-rich layer formed on high-strength bioactive glass-ceramic A–W. *J Biomed Mater Res* 1990;24:331–343.
12. Bonfield W, Grynpas MD, Tully AE, Bowman J, Abram J. Hydroxyapatite reinforced polyethylene — A mechanically compatible implant material for bone replacement. *Biomaterials* 1981;2:185–186.
13. Hench LL, Wilson J. Introduction. In: Hench LL, Wilson J, editor. *An Introduction to Bioceramics*. London: World Scientific; 1993; pp. 1–30.
14. Yamamuro T. A/W glass-ceramic: Clinical applications. In: Hench LL, Wilson J, editor. *An Introduction to Bioceramics*. London: World Scientific; 1993; pp. 89–104.
15. Rea SM, Bonfield W. Biocomposites for medical applications. *J Aust Ceram Soc* 2004;40:43–57.
16. Hench LL, Wilson J. *Clinical Performance of Skeletal Prostheses*. London: Chapman & Hall; 1996; p. 271.
17. Murphy CJ, Gole AM, Stone JW, Sisco PN, Alkilany AM, Goldsmith EC, Baxter SC. Gold nanoparticles in biology: Beyond toxicity to cellular imaging. *Acc Chem Res* 2008;41:1721–1730.
18. Ghosh P, Han G, De M, Kim CK, Rotello VM. Gold nanoparticles in delivery applications. *Adv Drug Del Rev* 2008;60:1307–1315.
19. Rosi NL, Giljohann DA, Thaxton CS, Lytton-Jean AKR, Han MS, Mirkin CA. Oligonucleotide-modified gold nanoparticles for intracellular gene regulation. *Science* 2006;312:1027–1030.
20. Minard-Basquin C, Kügler R, Matsuzawa NN, Yasuda A. Gold-nanoparticle-assisted oligonucleotide immobilisation for improved DNA detection. *IEE Proc Nanobiotechnol* 2005;152:97.
21. Yang K, Zhang S, Zhang G, Sun X, Lee ST, Liu, Z. Graphene in mice: Ultrahigh *in vivo* tumor uptake and efficient photothermal therapy. *Nano Lett* 2010;10:3318–3323.
22. Sun X, Liu Z, Welsher K, Robinson JT, Goodwin A, Zaric S, Dai H. Nano-graphene oxide for cellular imaging and drug delivery. *Nano Res* 2008;1:203–212.
23. Mohanty N, Berry V. Graphene-based single-bacterium resolution biodevice and DNA transistor: Interfacing graphene derivatives

with nanoscale and microscale biocomponents. *Nano Lett* 2008;8: 4469–4476.
24. Makadia HK, Siegel SJ. Polylactic-co-glycolic acid (PLGA) as biodegradable controlled drug delivery carrier. *Polymers* 2011;3:1377–1397.
25. Uhrich KE, Cannizzaro SM, Langer RS, Shakesheff KM. Polymeric systems for controlled drug release. *Chem Rev* 1999;99:3181–3198.
26. Schulz-Siegmund M, Hotzel R, Hoffmeister PG, Hacker MC. Synthesis, processing, and characterization of ceramic nanobiomaterials for biomedical applications. In: Sitharaman B, editor. *Nanobiomaterials Handbook*. Boca Raton, FL: CRC Press; 2011; pp. 3-1–3-30.
27. Raffa V, Vittorio O, Riggio C, Cuschieri A. Progress in nanotechnology for healthcare. *Min Invasive Ther Allied Technol* 2010;19:127–135.
28. Buzea C, Pacheco II, Robbie K. Nanomaterials and nanoparticles: Sources and toxicity. *Biointerphases* 2007;2:MR17–MR71.
29. Webster TJ, Ahn E. Nanostructured biomaterials for tissue engineering bone. In: Lee K, Kaplan D, editors. *Tissue Engineering II*. Berlin, Heidelberg: Springer; 2007; pp. 275–308.
30. Gutwein LG, Webster TJ. Increased viable osteoblast density in the presence of nanophase compared to conventional alumina and titania particles. *Biomaterials* 2004;25:4175–4183.
31. Kay S, Thapa A, Haberstroh KM, Webster TJ. Nanostructured polymer/nanophase ceramic composites enhance osteoblast and chondrocyte adhesion. *Tiss Eng* 2002;8:753–761.
32. Webster TJ, Schadler LS, Siegel RW, Bizios R. Mechanisms of enhanced osteoblast adhesion on nanophase alumina involve vitronectin. *Tiss Eng* 2001;7:291–301.
33. Webster TJ, Ergun C, Doremus RH, Siegel RW, Bizios R. Enhanced functions of osteoblasts on nanophase ceramics. *Biomaterials* 2000;21: 1803–1810.
34. Webster TJ, Ergun C, Doremus RH, Siegel RW, Bizios R. Specific proteins mediate enhanced osteoblast adhesion on nanophase ceramics. *J Biomed Mater Res* 2000;51:475–483.
35. Webster TJ, Siegel RW, Bizios R. Osteoblast adhesion on nanophase ceramics. *Biomaterials* 1999;20:1221–1227.
36. Webster TJ, Siegel RW, Bizios R. Design and evaluation of nanophase alumina for orthopaedic/dental applications. *Nanostruct Mater* 1999;12:983–986.
37. Kim K, Fisher JP. Nanoparticle technology in bone tissue engineering. *J Drug Target* 2007;15:241–252.

38. Ginebra MP, Traykova T, Planell JA. Calcium phosphate cements as bone drug delivery systems: A review. *J Control Rel* 2006;113:102–110.
39. Yu D, Wong J, Matsuda Y, Fox JL, Higuchi WI, Otsuka M. Self-setting hydroxyapatite cement: A novel skeletal drug-delivery system for antibiotics. *J Pharm Sci* 1992;81:529–531.
40. Richter V, Ruthendorf MV. On hardness and toughness of ultrafine and nanocrystalline hard materials. *Int J Refract Met Hard Mater* 1999;17:141–152.
41. Chokshi AH, Rosen A, Karch J, Gleiter H. On the validity of the Hall–Petch relationship in nanocrystalline materials. *Scr Mater* 1989;23: 1679–1683.
42. Mistry AS, Mikos AG, Jansen JA. Degradation and biocompatibility of a poly(propylene fumarate)-based/alumoxane nanocomposite for bone tissue engineering. *J Biomed Mater Res A* 2007;83A:940–953.
43. Wu D, Wu L, Zhang M. Rheology of multi-walled carbon nanotube/ poly(butylene terephthalate) composites. *J Polym Sci Part B Polym Phys* 2007;45:2239–2251.
44. Horch RA, Shahid N, Mistry AS, Timmer MD, Mikos AG, Barron AR. Nanoreinforcement of poly(propylene fumarate)-based networks with surface modified alumoxane nanoparticles for bone tissue engineering. *Biomacromolecules* 2004;5:1990–1998.
45. Ravagliolo A, Krajewski A. *Bioceramic: Material Properties Applications.* London: Chapman & Hall; 1992; p. 91.
46. Porter AE, Patel N, Skepper JN, Best SM, Bonfield W. Comparison of *in vivo* dissolution processes in hydroxyapatite and silicon-substituted hydroxyapatite bioceramics. *Biomaterials* 2003;24:4609–4620.
47. Ikeda N, Kawanabe K, Nakamura T. Quantitative comparison of osteoconduction of porous, dense A–W glass-ceramic and hydroxyapatite granules (effects of granule and pore sizes). *Biomaterials* 1999;20:1087–1095.
48. Horbett TA. Proteins: Structure, properties and adsorption to surfaces. In: Ratner BD, Hoffman AS, Schoe FJ, Lemmons JE, editor. *Biomaterials Science: An Introduction to Materials in Medicine.* New York: Academic Press; 1996; pp. 133–141.
49. Schakenraad JM. Cell: Their surfaces and interactions with materials. In: Ratner BD, Hoffman AS, Schoe FJ, Lemmons JE, editor. *Biomaterials Science: An Introduction to Materials in Medicine.* New York: Academic Press; 1996; pp. 141–147.

50. Oh S, Brammer KS, Li YS, Teng D, Engler AJ, Chien S, Jin S. Stem cell fate dictated solely by altered nanotube dimension. *Proc Natl Acad Sci USA* 2009;106:22130–22135.
51. Ahmad Z, Thian ES, Huang J, Edirisinghe MJ, Best SM, Jayasinghe SN, Bonfield W, Brooks RA, Rushton N. Deposition of nano-hydroxyapatite particles utilising direct and transitional electrohydrodynamic processes. *J Mater Sci Mater Med* 2008;19:3093–3104.
52. Wood JA, Ly I, Borjesson DL, Nealey PF, Russell P, Murphy CJ. The modulation of canine mesenchymal stem cells by nano-topographic cues. *Exp Cell Res* 2012;318:2438–2445.
53. Huang Y, Zhou G, Zheng L, Liu H, Niu X, Fan Y. Micro-/nano-sized hydroxyapatite directs differentiation of rat bone marrow derived mesenchymal stem cells towards an osteoblast lineage. *Nanoscale* 2012;4:2484–2490.
54. Webster TJ, Ergun C, Doremus RH, Siegel RW, Bizios R. Enhanced osteoclast-like cell functions on nanophase ceramics. *Biomaterials* 2001;22:1327–1333.
55. Liu X, Zhao X, Fu RKY, Ho JPY, Ding C, Chu PK. Plasma-treated nanostructured $TiO_2$ surface supporting biomimetic growth of apatite. *Biomaterials* 2005;26:6143–6150.
56. Wang G, Liu X, Zreiqat H, Ding C. Enhanced effects of nano-scale topography on the bioactivity and osteoblast behaviours of micron rough $ZrO_2$ coatings. *Colloids Surf B Biointerf* 2011;86:267–274.
57. Babister JC, Hails LA, Oreffo ROC, Davis SA, Mann S. The effect of pre-coating human bone marrow stromal cells with hydroxyapatite/amino acid nanoconjugates on osteogenesis. *Biomaterials* 2009;30:3174–3182.
58. Kretlow JD, Mikos AG. From material to tissue: Biomaterial development, scaffold fabrication, and tissue engineering. *AIChE J* 2008;54:3048–3067.
59. Gaharwar AK, Schexnailder PJ, Schmidt G. Nanocomposite polymer biomaterials for tissue repair of bone and cartilage. In: Sitharaman B, editor. *Nanobiomaterials Handbook*. Boca Raton, FL: CRC Press; 2011; pp. 24-1–24-20.
60. Riboldi SA, Sampaolesi M, Neuenschwander P, Cossu G, Mantero S. Electrospun degradable polyesterurethane membranes: Potential scaffolds for skeletal muscle tissue engineering. *Biomaterials* 2005;26:4606–4615.

61. Li WJ, Mauck RL, Tuan RS. Electrospun nanofibrous scaffolds: Production, characterization, and applications for tissue engineering and drug delivery. *J Biomed Nanotechnol* 2005;1:259–275.
62. Lee CH, Shin HJ, Cho IH, Kang YM, Kim IA, Park KD, Shin JW. Nanofiber alignment and direction of mechanical strain affect the ECM production of human ACL fibroblast. *Biomaterials* 2005;26: 1261–1270.
63. Kunzmann A, Andersson B, Thurnherr T, Krug H, Scheynius A, Fadeel B. Toxicology of engineered nanomaterials: Focus on biocompatibility, biodistribution and biodegradation. *Biochim Biophys Acta* 2011;1810: 361–373.

# Chapter 2

# Silica-based Nanoceramics

**Akiko Obata* and Toshihiro Kasuga†**

*Division of Advanced Ceramics,*
*Nagoya Institute of Technology, Gokiso-cho,*
*Showa-ku, Nagoya 466-8555, Japan*

* *obata.akiko@nitech.ac.jp*, † *t.kasuga.437@nitech.jp*

**Abstract**

Silica-based nanomaterials such as amorphous silica, mesoporous silica, and nanorods play an important role in biomedical applications, biochemistry, bioimaging, and biosensing. Thanks to the development of the sol–gel process, these nanomaterials feature uniform sizes/shapes and ordered pore sizes/structures, resulting in their unique properties. This chapter provides an overview of silica-based nanomaterials developed for biomedical applications, such as drug delivery and imaging, and their cytotoxicity. It also addresses their clinical use, such as in orthopaedics and dentistry, and their biocompatibility.

## 2.1 Introduction

Silicon exists as silicon dioxide (silica) and silicate in many minerals occurring in rocks, sand, and clay and finds application in many technical products such as glasses and cements. Its chemistry shows several similarities to that of carbon as well as some striking differences.[1] Silicon and carbon form long-chain polymers containing alternating silicon–oxygen or carbon–oxygen backbones, respectively. Silicon-based materials comprise weaker Si–Si and Si–O bonds compared with the corresponding C–C and C–O bonds, resulting in their greater biodegradability and thus greater biocompatibility than their carbon-based counterparts.[2] Despite presenting the same number of outer shell electrons, silicon and carbon exhibit different structural, chemical, and physical properties.

Silica-based biomaterials have widely been used in drug-delivery system,[3,4] biological sensing,[5,6] imaging,[7,8] orthopaedics and dental filling,[9] cementing,[10] and metallic material coatings.[11] Their biocompatibility and safety have been investigated extensively because of their *in vivo* implementation. These properties depend on the composition and size. In particular, a nanometric size provides novel and significantly improved physical, chemical, and biological properties to nanomaterials but contributes to their cell and tissue toxicity in the body.

Numerous studies have led to chemical surface modification approaches to reduce this toxicity. The sol–gel process is the most common method for the fabrication of silica-based nanomaterials, particles, and coatings. In addition to accurate size and morphology control, this process achieves a tailored surface functionalisation by manipulating several conditions, such as catalyst and temperature. Drug-delivery systems rely on the nanoparticle porous structure, whereas imaging applications require controlled diameter and shape.

Bone disorders have emerged as a significant concern because of the increase in the median age of the population, prompting a growing interest in the effects of silicon and silicate ions on bone regeneration. Silica-based biomaterials, such as glasses and cements, have shown excellent bioactivity and, thus, a strong ability to bond to natural bone, making them attractive for bone and tooth regeneration.[9] Silicon has been proved to be an essential trace element in animal nutrition, playing an important role in bone and cartilage.[12] Specifically, silicon originating from dietary intake and silicate ions present in these biomaterials effectively contribute to bone and cartilage formation. The released silicate ions stimulate osteogenic cell functions upon dissolution, enhancing bone formation.[13]

## 2.2 Silica-based Nanomaterials

The implementation of silica-based nanomaterials in a wide variety of industrial and consumer applications, such as diet, cosmetic, and medical products, has been investigated for many decades. In particular, their application to drug delivery and biological imaging has attracted tremendous attention.

As mentioned above, these nanomaterials, including mesoporous silica nanoparticles (MSNs), are typically fabricated by the sol–gel process using triethoxysilane (TEOS) at low temperature and pressure. This process, comprising a two-step inorganic polycondensation (Eqs. (2.1)–(2.3)), achieves an accurate control of morphology, pore size and structure, and surface functionalisation.

**Hydrolysis:**

$$\mathrm{Si(OR)_4} + x\mathrm{H_2O} \rightarrow \mathrm{(RO)_{3-x}\text{–}Si\text{–}(OH)_x} + x\mathrm{ROH} \tag{2.1}$$

**Condensation:**

$$\mathrm{\text{–}Si\text{–}OR + \text{–}Si\text{–}OH \rightarrow \text{–}Si\text{–}O\text{–}Si\text{–} + ROH} \tag{2.2}$$

$$\mathrm{\text{–}Si\text{–}OH + \text{–}Si\text{–}OH \rightarrow \text{–}Si\text{–}O\text{–}Si\text{–} + H_2O} \tag{2.3}$$

These characteristics can be rationally designed by manipulating the water/silane ratio, catalyst, temperature, and solvent type.

According to the ISO Core Terms, nanomaterials are industrial materials intentionally produced, manufactured, or engineered to exhibit unique properties or specific composition at the nanoscale, which ranges from approximately 1 to 100 nm.[14] They have been generated in various shapes, including particles, fibres, plates, and rods. Their physico-chemical characteristics, such as size, surface charge, and shape, have reportedly elicited different biological responses, including cellular uptake, distribution, elimination, haemolytic activity, genotoxicity, and carcinogenicity. In particular, their toxicity has become a great concern concomitant with their incorporation into an increasing number of consumer products.

Primary nanoobjects, e.g., particles, belong to the nanoscale but others are composed of their agglomerates. Therefore, existing studies have generally addressed these primary structures rather than their agglomerates.[14] Nonetheless, micrometric agglomerates may be broken down into the size of primary nanomaterials in the body. Their formation depends on nanomaterial size and surrounding liquid properties, such as pH, which affect other toxicologically relevant characteristics, such as the surface area and particle number. This section focuses on silica-based nanomaterials for drug delivery and imaging.

### 2.2.1 Drug delivery

Several nanostructured porous materials have been developed to obtain controlled functional drug-loading systems. Used as vehicles, these materials are expected to enhance the transport of drugs and bioactive substances because of their nanostructure. In particular, silica-based materials are considered as good vehicle candidates because of (1) *an ordered pore network*, which is very homogeneous in size and facilitates fine control of drug loading and release kinetics, (2) *a high pore volume*, which allows for the required amount of pharmaceuticals to be hosted, (3) *a high surface area*, which implies a high capacity for drug adsorption, and (4) *a silanol-bearing surface* that can be functionalised to better control drug loading and release.[15]

Tang *et al.*[4] reviewed the fabrication, properties, and application of MSNs for drug delivery. According to International Union of Pure and Applied Chemistry (IUPAC), mesoporous materials contain 2–50-nm-diameter pores. Most MSNs developed for drug delivery, such as ordered mesoporous and hollow or rattle-type silica, display uniform pore sizes and ordered pore structures.

In addition to drug delivery, ordered mesoporous silica finds its way into molecular separation, catalysis, and sensing.[16] These

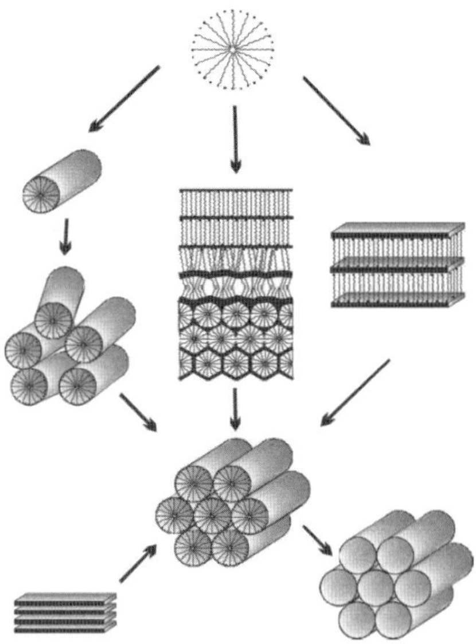

Figure 2.1 General pathway for the formation of ordered mesoporous materials. Reprinted with permission from Ref. 16. Copyright 2005 American Chemical Society.

materials possess high specific surface areas and pore volumes as well as accurately ordered pore structures featuring uniform mesopores in diameters adjustable from about 1.5 to 30 nm (Fig. 2.1). The mesopores assist protein adsorption and retain their functions.

The most studied ordered mesoporous silica, Mobil Composition of Matter No. 41 (MCM-41), comprises honeycomb arrays of non-intersecting uniformly sized channels presenting diameters ranging from 1.5 to 10 nm. Studies on biomedical applications of MSNs have multiplied[4,16] since the first application of MCM-41 to drug-delivery systems by Vallet-Regí *et al.*[17] Their drug loading and release ability are controlled by optimising parameters, such as pore size and surface area. For example, the large-pore MCM-41 prepared using hexadecyltrimethylammonium bromide released 68% of the loaded ibuprofen after 24 h in simulated body fluid, whereas its small-pore analogue obtained using dodecyltrimethylammonium bromide delivered 55% of the drug.[15] Therefore, the MCM-41 pore size regulates molecular sieving function and drug-release rate.

Hollow MSNs have recently become the subject of increasing interest, which has led to numerous new fabrication methods. These nanoparticles are prepared by a dual template method combining a hard or soft template to generate a hollow interior with a soft template to produce mesopores in the shell.

Tang *et al.*[4] categorised the synthetic methods for hollow MSNs into soft template, selective etching, and self-template methods. In soft template methods, surfactants form a soft template, whereas dual or multi-surfactants produce a complex template to simultaneously build the mesoporous shell and its hollow interior. The selective etching strategy combines a pure silica framework and hybrid organic–inorganic networks presenting different compositions and structures, which may induce different stabilities using various etching agents or under exceptional temperature or pH conditions. The self-template method generates hollow MSNs without extra templates.

### 2.2.2 Imaging

Quantum dots (QDs) exhibit unique photophysical properties which are applicable to photonics and biochemistry. Their emission spectra are tunable by varying the primary particle size or composition, which makes them powerful light sources for next-generation biotechnological devices. For example, they have been utilised for the extended imaging of cells as well as subcellular organelle structures and functions.

Biju *et al.*[7] reviewed extracellular and intracellular imaging using QDs. Fluorescent dyes are most commonly used for biological staining and labelling but suffer from several limitations, such as narrow excitation bands, small Stokes shifts, broad fluorescence bands, and photobleaching. On the other hand, QDs benefit from size-tunable absorption and emission spectra in the visible and near-infrared (NIR) regions, narrow emission and broad absorption bands, and large multiphoton absorption cross-sections.

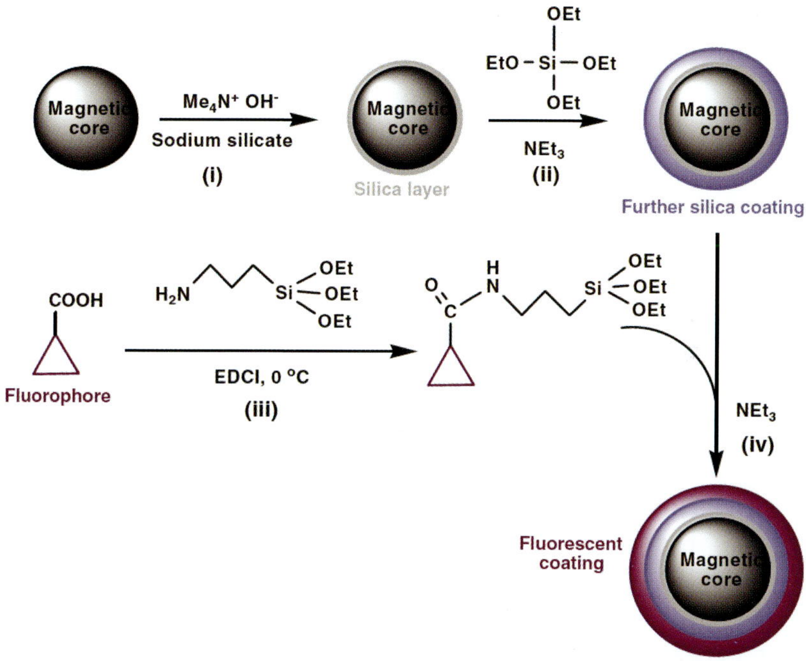


Figure 2.2 Preparation of fluorescently labelled silica-coated magnetic nanocomposites. (i) Initial optional coating with sodium silicate; (ii) base-catalysed condensation of TEOS on the nanoparticle surface; (iii) covalent attachment of carboxyl fluorophore to 3-aminopropyltriethoxysilane via 1-ethyl-3-(3-dimethylaminopropyl)carbodiimide (EDC) coupling; (iv) condensation of silane-modified fluorophores onto a silica-coated magnetic particle. Reprinted with permission from Ref. 8.

Silica coatings significantly improve the stability and compatibility of QDs in aqueous media. Correa-Duarte *et al.*[18] prepared silica shells on CdS QDs and discovered the suppression of their photooxidation-induced destabilisation in the presence of a homogeneous, microporous silica shell. They suggested that the shell reduced the QD access of $O_2$, causing this suppression.

Li *et al.*[19] revealed that the encapsulation of CdSe QDs by silica-coated gold nanorods (GNRs_$SiO_2$) enhanced their fluorescence and optimised their emission by modifying the silica coating thickness. The QDs were efficiently functionalised with anti-CEA8 and conjugated with GNRs and GNRs_$SiO_2$ taken up by the cancer cells. GNRs_$SiO_2$ showed an especially enhanced fluorescence.

Corr *et al.*[8] reviewed magnetofluorescent nanocomposite materials, such as fluorophore-encapsulated, silica-coated magnetic nanoparticles (Fig. 2.2). Magnetic nanoparticles can be brought to a target site in the body using external magnetic fields, which makes them powerful components of site-specific drug-delivery systems. In addition, they are expected to heat up when subjected to an external magnetic AC field, which opens doors for hyperthermic cancer treatment.

Silica has been deposited on magnetic particles to produce fluorescent-magnetic nanocomposites. The silica coating acts as an interlayer with the following features. (1) It provides an effective barrier to any fluorophore quenching by the magnetic cores. (2) It facilitates the direct encapsulation of fluorescent dyes or QDs because of its relative inertness and optical transparency. (3) The surface is easily functionalised, which enables the chemical binding of various fluorescent and biological species. (4) The coating may reduce any potential toxic effects of the bare nanoparticles.[8]

## 2.3 Biocompatibility of Silica-based Nanomaterials

Numerous *in vitro* and *in vivo* studies have examined the biocompatibility of silica-based nanomaterials, especially cellular uptake, haemolytic activity, distribution and elimination from the body, repeated dose toxicity, genotoxicity, and carcinogenicity.[14] These cellular and physical responses to the materials relate to size, surface properties, shape, and structure (Figure 2.3). Silica is generally considered biocompatible because of its low toxicity. However, its nanostructured derivatives need to be reassessed because of their specific chemical and physical properties.

### 2.3.1 Cellular uptake

Many silica-based nanomaterials are used for cell labelling, gene delivery, and intracellular drug delivery, which rely on cellular uptake. In this process, the nanomaterials adhere to the cell

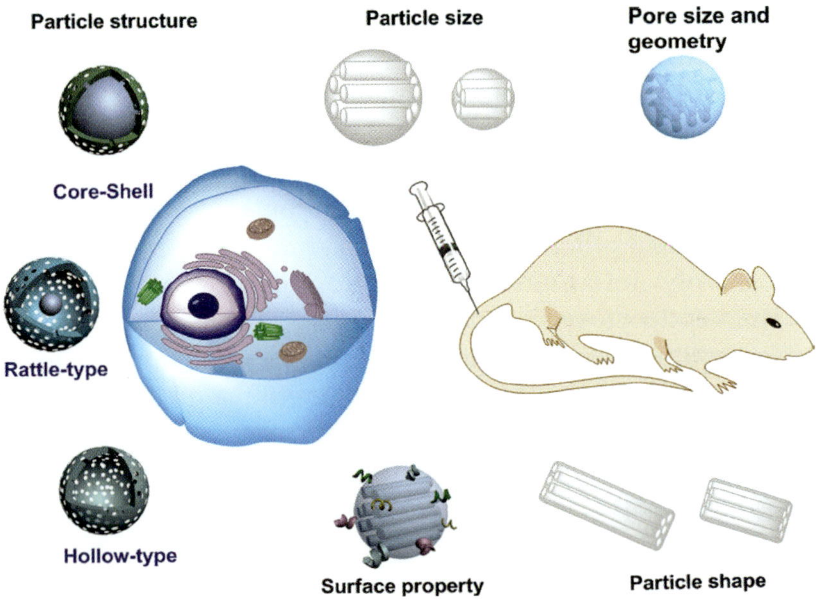


Figure 2.3 Schematic illustration of biocompatibility and biotranslocation of MSNs. Reprinted with permission from Ref. 4.

membrane through specific or non-specific interactions before being engulfed by the membrane and taken up through several different endocytic pathways, such as pinocytosis, phagocytosis, and receptor-mediated endocytosis. Their cellular uptake occurs upon exposure to cells depending on their size. Microparticles (>0.5 $\mu$m) are usually ingested through phagocytosis, and smaller objects (<0.2 $\mu$m) undergo pinocytosis.[20]

Zhang *et al.*[21] studied the interactions of 80- and 500-nm-diameter silica nanoparticles with human dermal fibroblasts by monitoring the cellular uptake of the particles at different time intervals. The uptake amount of 500-nm-diameter particles was the same as that of their 80-nm-diameter analogues after 4 h but surpassed it after 24 h, showing that the uptake behaviour depends on the nanoparticle size.

Conversely, the researchers found a different size dependence for other cells. HeLa cells took up more 50-nm-diameter gold nanoparticles than 14- and 74-nm-diameter ones,[22] whereas Caco-2 cells absorbed fewer 50-nm-diameter poly(lactic-co-glycolic acid)

(PLGA) particles than 200- and 500-nm-diameter ones.[23] This difference in cellular uptake behaviour may imply diverse mechanisms.

Silica nanoparticles with diameter ranging from 80- to 500 nm followed different uptake mechanisms. The 80-nm-diameter nanoparticles predominantly underwent micropinocytosis and clatherin-mediated endocytosis in addition to caveolae-mediated uptake. On the other hand, their 500-nm-diameter counterparts were mainly internalised by micropinocytosis and clathrin-mediated endocytosis.[21]

Costantini *et al.*[24] compared the phagocytosis and toxicity of amorphous and crystalline silica particles in the presence of alveolar macrophages (MH-S). Despite their identical molecular formula, crystalline silica exhibits higher inhalation toxicity than its amorphous analogue. In addition to its naturally occurring and synthetic sources, amorphous silica is generated as a by-product of power stations and metallurgical processing. Synthetic amorphous silica finds applications in pharmaceuticals, paints, cosmetics, and food as additives and fillers.[25]

Culture tests using MH-S cells demonstrated that amorphous and crystalline silica particles induced high cell death rates. All silica-treated cells died in 24 h. RAW264.7 macrophage cells showed similar responses to both silica types. This phenomenon was also observed for different particle sizes of amorphous silica. On the other hand, no toxicity was detected when the cells were cultured in the presence of particles in Transwell chambers (indirect treatment). Therefore, cell death resulted from direct interactions between cells and silica particles.

Macrophages have proved to be especially sensitive to silica particle toxicity. Table 2.1 shows the death and particle uptake rates for various cell types. All cells internalised the amorphous silica particles within 16 h, but only MH-S cells showed death rates,[24] indicating that the cell type used for cellular uptake and cytotoxicity tests is an important factor influencing conclusions on the biocompatibility of silica-based materials.

Chung *et al.*[26] evaluated the effects of MSN surface charge on cellular uptake. The surface charge was varied from a zeta potential of –4.90 (as prepared) to +19.0 by changing the degree of surface

Table 2.1 Cell death and particle uptake rates of various types of cell lines. Reprinted with permission from Ref. 24.

| | | Cell Death (% Membrane Permeability*) | | | Particle Uptake‡ | |
|---|---|---|---|---|---|---|
| Cell line | Description | Control (cell alone) | Crystalline silica | Amorphous silica | % Internalized Particles† | Particles/Cell |
| MHS (mouse) | Macrophage | 17 | 100 | 100 | 81 | 4.2 |
| B16F1 (mouse) | skin melanoma | 0.45 | 0 | 0.15 | 85 | 8 |
| Cos7 (monkey) | kidney fibroblast | 0 | 1.3 | 5 | 79 | 6 |
| HeLa (human) | cervical epithelial | 0 | 0 | 0 | 39 | 9 |
| MDCK (dog) | kidney epithelial | 0 | 0 | 0 | 93 | 6.9 |
| MTLN3 (mouse) | mammaty | 0 | 0 | 0 | 98 | 5.1 |
| NIH3T3 (mouse) | embryo fibroblast | 0 | 0 | 0 | 59 | 8.3 |

Cells treated with crystalline silica or 3 mm amorphous silica [25 $\mu g/cm^2$].
*Cell death induced at 24 h, determined by membrane permeability (PI positive).
†% internalized equals the number of internalized particles divided by total number of cell associated particles.
‡Uptake measurements at 16 h after silica exposure.
doi:10.1371/journal.pone.0014647.t002.

modification using *N*-trimethoxysilylpropyl-*N*,*N*,*N*-trimethylammonium chloride. The particle uptake by 3T3-L1 cells increased with increasing positive particle surface charge. In contrast, internalisation rates did not change significantly for human mesenchymal stem cells (hMSCs) when particle surface charges varied.

This uptake improvement observed with increasing surface charge agreed with other reports, showing a clear correlation between polystyrene nanoparticle surface charge and fluorescence intensity in different cell lines (HeLa, Jurkat, and KG1s cells), with the exception of hMSCs.[27] In addition, they also mentioned that MSNs displayed a high cellular labelling efficiency. The positive surface charge had negligible effect on the uptake efficiency of MSNs.

### 2.3.2 Haemolysis

Silica-based materials comprising amorphous particles cause mammalian red blood cell haemolysis. Proposed haemolysis pathways include the cell membrane protein denaturation through electrostatic interactions with silicate, the generation of reactive oxygen species by the silica surface, and the high binding affinity of silicate for tetra-alkylammonium groups that are abundant in red blood cell membranes.[28] Nonetheless, membrane damage mainly occurs through the denaturation of cell membrane proteins by proton-donating silanol groups.

Pandurangi *et al.*[29] found a strong correlation between haemolysis and the number of silanol groups on amorphous silica particles. They examined the haemolytic activities of Min-U-Sil particles, which form a widely used crystalline $\alpha$-quartz-type dust, and amorphous and fumed Cab-O-Sil particles exhibiting a very large specific surface area. The number of silanol groups on these particles was measured by Fourier transform infrared spectroscopy at various heat-treatment temperatures. The untreated Cab-O-Sil particle surface bore more silanol groups than its Min-U-Sil equivalent and showed about an order of magnitude higher haemolytic activity. The haemolytic activity changed with increasing temperature, indicating its dependence on the number of silanol groups on the particle surfaces. These measurements revealed a positive correlation between the haemolytic activity of the silica particles and the surface silanol band intensity at 3745 $cm^{-1}$ for both particles.

Slowing *et al.*[28] showed that haemolysis depends on the concentration of negatively charged silanol groups accessible to red blood cell membranes by comparing the haemolytic activities of MSNs and amorphous silica particles. Direct- and cross-polarisation magic-angle spinning $^{29}Si$ NMR spectra demonstrated that MSNs exhibited fewer surface silanol groups than amorphous silica particles. The $(Q^2 + Q^3)/Q^4$ ratio, in which a $Q^n$ unit corresponds to a silicon atom with $n$ bridging oxygen atoms (–Si–O–Si–) and 4-$n$ non-bridging oxygen atoms, amounted to 0.8 and 0.4 for MSNs and amorphous silica, respectively. On the other hand, haemolysis tests demonstrated that amorphous silica particles presented higher haemolytic activity than MSNs. As a consequence, the haemolytic properties of these materials are not proportional to the total number of silanol groups because most of these groups are located on the MSN interior surface instead of its exterior.

They also revealed that the haemolytic activity was highly specific to silanol groups by determining the haemolytic activities of carboxy- and sulfonate-functionalised amorphous silica particles. All particles exhibited almost the same zeta potentials, but their haemolytic activity decreased from 45% to less than 1% by functionalisation with carboxy and sulfonate groups, indicating that haemolysis only depended on silanol groups.

### 2.3.3 Distribution and elimination

The inhalation of silica-based microparticles by humans has been linked to the pulmonary disease silicosis depending on their shape.[30] Silica-based nanomaterials may elicit harmful toxicological effects because of their small size and present dramatically different biological effects from their microscale counterparts *in vivo*. In addition, *in vitro* results are difficult to extrapolate into *in vivo* effects, making detailed *in vivo* investigations of silica-based nanomaterial properties necessary.

Tissue distribution and excretion of injected nanoparticles vary depending on the particle size. Nabeshi *et al.*[31] examined the skin penetration, cellular localisation, and cytotoxic effects of well-dispersed amorphous silica nanoparticles measuring 70 nm, 300 nm, and 1000 nm in diameter. After 28 days of dermal administration, the 70-nm-diameter nanoparticles entered the skin, the lymph nodes, and the liver of BALB/c mice in addition to the cerebral cortex and the hippocampus. They were also localised in the nuclei of skin and lymph node cells. In addition, the transdermal application of the nanoparticles induced cellular damage to the skin. These phenomena were only observed for 70-nm-diameter nanoparticles.

*In vitro* tests involving the same nanoparticles using HaCaT cells showed that smaller nanoparticles (70 nm and 300 nm) inhibited cell growth more strongly than their larger analogues. In addition, mutagenicity tests using *Salmonella typhimurium* (Ames test) revealed that the nanoparticle mutagenicity increased with decreasing particle size. Therefore, biological effects also depend on the particle size. This may stem from the genotoxicity induced by the accumulation of smaller nanoparticles in cell nuclei. Injected silica nanoparticles have caused pregnancy complications in mice depending on their size.[32]

As mentioned above, residues from injected silica nanoparticles in the body are expected to cause acute toxicity. Therefore, an effective nanoparticle clearance may prevent toxicity induced by residual ions and foreign materials in the body. He *et al.*[33] examined the distribution and the urinary excretion of three types of 45-nm-sized OH–, COOH–, and polyethylene glycol (PEG)-modified silica nanoparticles in mice by optical imaging. Intravenously injected nanoparticles were cleared from the circulatory system but existed inside organs. Surface modifications impacted biological reactions. The PEG–silica nanoparticles exhibited the longest blood circulation times and lowest uptake by the liver, indicating that surface modification may effectively eliminate injected nanoparticles from the body.

Most porous particles are accumulated in the liver and spleen. Huang *et al.*[34] discovered that clearance depends on the nanoparticle shape. Short rod-shaped silica nanoparticles were trapped in the liver, whereas their long analogues stayed in the spleen. Therefore, shorter nanoparticles were cleared out more quickly.

## 2.4 Silica-based Materials for Bone and Tooth Regeneration

Bioceramics play an important role in the repair and reconstruction of diseased or damaged parts of the musculoskeletal system. They are roughly classified as bioinert and bioactive materials according to their reactivity toward biological tissue and cells. Bioinert materials minimise the immune response to a foreign body while achieving a suitable combination of physical properties to match those of the replaced tissue. On the other hand, bioactive materials elicit a controlled action and reaction in the physiological environment.[35–38] Various silica-based bioactive materials have been developed and applied to orthopaedic and dental fields. Most of these materials are bioresorbable and release soluble silica and calcium ions into their surroundings. The released soluble silica contributes to the excellent osteogenic properties of the materials,[39,40] whereas silicate ions facilitate bone tissue formation and calcification.[12]

### 2.4.1 Bioactive glasses

Hench, from the University of Florida, invented the first bioactive glass, known as 45S5 or Bioglass®, in 1969. This material, which consisted of 46.1 mol.% $SiO_2$, 24.4 mol.% $Na_2O$, 26.9 mol.% CaO, and 2.6 mol.% $P_2O_5$, formed a bond with bone.[9] Its development launched the field of bioactive ceramics. Bioactive glasses, glass-ceramics, and ceramics have led to several materials and products.

Jones[9] reviewed bioactive glass materials and products derived from 45S5. Existing clinical products include the middle ear prosthesis (MEP®), the Endosseous Ridge Maintenance Implant (ERMI®) for tooth root repair, the PerioGlas® for bone graft, the NovaBone® for the orthopaedic bone grafting of non-load-bearing sites, the

Biogran® for the bone graft used in jaw bone defect regeneration, and the NovaMin® for toothpaste.

### 2.4.2 Silica-mediated hydroxycarbonate apatite formation

The 45S5 surface exhibits a biologically active hydroxycarbonate apatite (HCA) layer that provides a bonding interface with tissues. Chemically and structurally equivalent to the mineral phase in bone, the HCA layer can incorporate collagen fibrils when growing on 45S5 surfaces. Its formation relies on the silica present in 45S5 according to the following mechanism[38]:

1. Rapid exchange of $Na^+$ with solution-phase $H^+$ or $H_3O^+$.
2. Dissolution of $SiO_2$ as $Si(OH)_4$ resulting from Si–O–Si bond breaking and Si–OH formation at the glass–solution interface.
3. Condensation and repolymerisation of a $SiO_2$-rich layer on the surface depleted in alkali and alkali earth cations.
4. Migration of $Ca^{2+}$ and $PO_4^{3-}$ groups to the surface through the $SiO_2$-rich layer, generating a $CaO–P_2O_5$-rich film over this layer and, subsequently, amorphous $CaO–P_2O_5$ film by incorporating soluble calcium and phosphates from the solution.
5. Crystallisation of the amorphous $CaO–P_2O_5$ film into a mixed hydroxyl, carbonate apatite layer by incorporation of $OH^-$ and $CO_3^{2-}$ ions from solution.

### 2.4.3 Effects of soluble silica on osteogenesis

Soluble silica and other inorganic ions released from glass also play a central role for achieving direct bonding with bone. Hoppe *et al.*[13] reviewed the biological response elicited by inorganic ions derived from bioactive glasses and glass-ceramics. These ions enhance osteogenesis, angiogenesis, and antibacterial activity. In particular, the effects of silicate ions on osteogenic cell functions have been extensively investigated.

Xynos *et al.*[39,41] showed that ionic species, such as soluble silica and calcium ions, increased human osteoblast proliferation by gene activation. Reffitt *et al.*[42] investigated the effects of orthosilicic acid on collagen type I (COL I) synthesis and osteoblastic differentiation using four different types of cells, such as primary cells, and found that the acid enhances the COL I synthesis in all treated cells.

### 2.4.4 Silicate ions dissolved from bioactive glass *in vivo*

Most bioactive glasses implanted in the body release silicate and other inorganic ions that may induce pathogenesis in the host. Kawanabe *et al.*[43,44] administered bioactive material powders including 45S5 in mice and rats (dose: 1 or 5 mg/g body weight) by intramuscular, subcutaneous, and intraperitoneal injections and evaluated nephrotoxicity induced by the powder-derived ionic products. Intramuscular and subcutaneous injections did not cause any adverse outcome. In contrast, silicon accumulated in renal tissue when silicate compounds were injected by the intraperitoneal route, resulting in severe effects. The researchers suggested that these effects only occurred by intraperitoneal injection because the silicate concentration increased until polymerisation proceeded as a result of the abundant blood circulation and absorptive peritoneum surface. These effects also stemmed from the relative hydration and fluid volume available for mineral leaching.

Lai *et al.*[45,46] implanted 45S5 granules in rabbit tibiae or muscles and investigated the pathway of silicate ions released *in vivo*. The dose was 1500 mg/4 kg body weight to represent a relevant dose for a human clinical case, in which the granules would act as a bone graft substitute. Silicon concentrations in urine were well below saturation (180 ppm) over 24 weeks after the implantation. Silicon concentrations were measured in various fluids and tissues, such as blood, kidney, liver lung, lymph node, spleen, and tibia/muscle. The only significant difference between the implanted and control groups was found in the tibia/muscle which is the site of implantation, suggesting that the resorbed silica is harmlessly excreted in its soluble form through the urine.

### 2.4.5 Sol–gel-derived 70S30C-type bioactive glass

Sol–gel-derived glasses present simpler compositions than their melt-derived analogues, such as 45S5. Their mesoporous texture, which is inherent to the sol–gel process, increases their specific surface area, which is expected to enhance their bioactivity and resorbability. Jones *et al.* developed a binary $SiO_2$–CaO glass (70 mol.% $SiO_2$, 30 mol.% CaO, 70S30C)[47] and fabricated three-dimensional (3D) bioactive scaffolds exhibiting hierarchically interconnected pore morphologies similar to trabecular bone by a foaming method.[48]

An optimal scaffold for bone tissue engineering applications needs to template *in vitro* and *in vivo* bone growth. An interconnected macroporous network with a modal interconnected pore diameter exceeding 100 $\mu$m may enable cell migration, tissue ingrowth, and eventually vascularisation. The obtained 70S30C-containing 3D scaffold showed a hierarchical structure, which comprised macropores larger than 500 $\mu$m and was connected by pore windows exceeding 100 $\mu$m in diameter, as well as a mesoporous texture presenting 10–20-nm pore diameters. When sintered at 800°C for 2 h, the scaffolds also fulfilled the criteria for an ideal scaffold for bone tissue engineering, including in terms of their mechanical properties. Their compressive strength amounted to 2.3 MPa, which is in the same range as for trabecular bone (2–12 MPa).

The 70S30C scaffolds induced the osteogenic differentiation of primary human osteoblasts.[49] The cells were cultured on the scaffolds and their differentiation levels were evaluated by measuring their alkaline phosphatase (ALP) activity and COL I protein production. These cultured cells underwent differentiation even in the absence of supplementary factors, such as ascorbic acid, $\beta$-glycerophosphate, and dexamethasone. This means that the scaffold itself possesses the ability to induce the osteogenic differentiation of the osteoblasts, fulfilling many of the criteria for an ideal bone tissue engineering scaffold.

### 2.4.6 Silicon-substituted hydroxyapatite

Hydroxyapatite [$Ca_{10}(PO_4)_6(OH)_2$, HA] has been applied as a bone graft material in medicine and dentistry. However, it presents a lower reactivity toward existing bone and contacts bone at slower rates compared to bioactive glasses and glass-ceramics. Silicon-substituted HA (Si-HA) was developed to solve these problems and exhibited enhanced properties *in vitro* and *in vivo*. Bonfield and coworkers[50] found that Si-HA granular implants displayed significantly greater bone apposition on their surface than their HA counterparts. The silicon dissolution in the bulk regions of the implant followed the same order as the bone apposition percentage at the ceramic surface *in vivo*, i.e., 1.5 > 0.8 > 0 (wt.% of Si).[51,52] These findings indicate that the incorporation of silicate ions into HA increases Si-HA dissolution rates, consistent with the accelerated biological apatite precipitation around the implant.

Si-HA enhances osteoblast functions, which contributes to its higher bone apposition level *in vivo* compared to HA.[53] Human osteoblasts exhibited significantly higher proliferation and ALP production on Si-HA discs prepared using the granules than their HA equivalents. These effects of silicon on osteoblasts were also observed for Si-HA thin coatings prepared by magnetron co-sputtering. Thin coatings containing more silicon than the aforementioned granules were developed as bioactive coatings for metallic implants, such as titanium.[54,55] Human osteoblast-like cells were cultured on titanium plates without or with Si-HA coating containing 0.8 wt.%, 2.2 wt.%, and 4.9 wt.% of Si. These cells showed improved adhesion and mineralisation on Si-HA coatings with increasing silicon content.

Aizawa *et al.*[56] developed Si-HA fibre scaffolds for bone tissue engineering. The Si-HA fibres were prepared by homogeneous precipitation, mixed with carbon beads, and compacted by pressing. The resulting monoliths were fired at 1300°C for 5 h to generate the porous fibre scaffolds. These scaffolds showed an enhanced mouse osteoblast proliferation (MC3T3-E1 cells) compared to their HA equivalents.[57] They released silicate ions in buffer solutions, which is expected to promote cell proliferation.

### 2.4.7 Siloxane-containing vaterite

The calcium carbonate polymorph vaterite exhibits higher dissolution ability than calcite and aragonite. Spherical vaterite particles

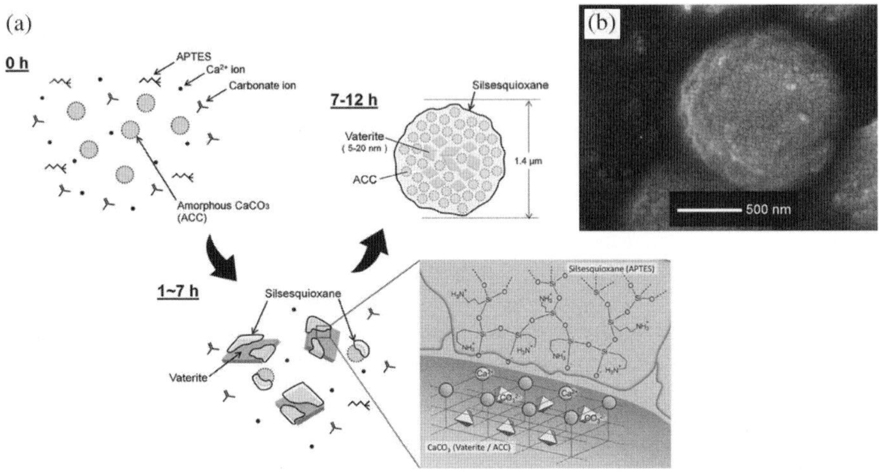


Figure 2.4 (a) Proposed steps involved in SiV particle formation. (b) SEM micrographs of SiV particles. Reproduced from Ref. 58 with permission from The Royal Society of Chemistry.

measuring about 500 nm in diameter are synthesised by a carbonation process. Kasuga and coworkers[58,59] generated 1.5-$\mu$m-diameter siloxane-containing vaterite (SiV) particles by adding aminopropyltriethoxysilane (APTES) to the vaterite precursor (Fig. 2.4). SiV particles containing 2.6 wt.% silicon presented uniform, spherical morphologies resulting from the agglomeration of its primary particles. They comprised a crystalline vaterite particle mixture where APTES molecules condense in the vicinity, amorphous calcium carbonate adsorbed by APTES silanol groups, and APTES-derived silsesquioxane.[58] They released calcium and soluble silica species by dissolving in an aqueous solution. The release rate of calcium ions was tuned by enhancing the crystalline plane orientation in the SiV.[59]

Composites consisting of SiV particles and biodegradable polymers, such as poly(l-lactic acid) (PLLA) or PLGA, have been developed by the same research group.[60–64] Non-woven[60,62] and cotton-wool-like fibrous composites[63,64] prepared by electrospinning benefit from easy handling for bone-filling surgery and the ability to release calcium and soluble silica species, inducing cell migration *in vitro* and enhancing mineral tissue formation *in vivo*.

The fibre diameter was controlled by regulating the electrospinning conditions, such as the composite solution viscosity. SiV/PLLA or SiV/PLGA composite fibres measuring several hundreds of nanometres to several tens of micrometres in diameter were prepared. Electrospun fibres exhibiting diameters larger than 10 $\mu$m enabled cell migration within their porous structures.[65] The composite fibres induced biological apatite precipitation in simulated body fluid. Culture tests using MC3T3-E1 cells demonstrated a faster cell proliferation and differentiation on SiV/PLLA composites than on their vaterite/PLLA analogues.[61] Therefore, non-woven and cotton-wool-like SiV/PLLA and SiV/PLGA composites may be useful for enhancing bone regeneration.

## 2.5 Concluding Remarks

Silica-based materials have attracted great interest in many fields of science, such as pharmaceutics, orthopaedics, and dentistry, because of the unique properties of silicon (silica). In particular, nanomaterials have emerged as desirable materials in nanobiotechnology and nanomedicine, prompting an exponential increase in research on these materials for biomedical applications in the last decade. In general, silica-based materials are not considered harmful to humans. However, the chemical and physical effects of nanomaterials require attention because of their nanosize.

## References

1. Müller WEG, Grachev MA. *Biosilica in Evolution, Morphogenesis, and Nanobiotechnology*. Berlin, Germany: Springer; 2009.
2. Jaganathan H, Godin B. Biocompatibility assessment of Si-based nano- and micro-particles. *Adv Drug Deliv Rev* 2012;64(15):1800–1819.
3. Shahbazi M-A, Herranz B, Santos HA. Nanostructured porous Si-based nanoparticles for targeted drug delivery. *Biomatter* 2012;2(4): 296–312.

4. Tang F, Li L, Chen D. Mesoporous silica nanoparticles: Synthesis, biocompatibility and drug delivery. *Adv Mater* 2012;24(12):1504–1534.
5. Zhao X, Tapec-Dytioco R, Tan W. Ultrasensitive DNA detection using highly fluorescent bioconjugated nanoparticles. *J Am Chem Soc* 2003;125(38):11474–11475.
6. Santra S, Zhang P, Wang K, Tapec R, Tan W. Conjugation of biomolecules with luminophore-doped silica nanoparticles for photostable biomarkers. *Anal Chem* 2001;73(20):4988–4993.
7. Biju V, Itoh T, Ishikawa M. Delivering quantum dots to cells: Bioconjugated quantum dots for targeted and nonspecific extracellular and intracellular imaging. *Chem Soc Rev* 2010;39(8):3031–3056.
8. Corr S, Rakovich Y, Gun'ko Y. Multifunctional magnetic-fluorescent nanocomposites for biomedical applications. *Nanoscale Res Lett* 2008;3(3):87–104.
9. Jones JR. Review of bioactive glass: From Hench to hybrids. *Acta Biomater* 2013;9(1):4457–4486.
10. Shie M-Y, Ding S-J, Chang H-C. The role of silicon in osteoblast-like cell proliferation and apoptosis. *Acta Biomater* 2011;7(6):2604–2614.
11. Stan GE, Popa AC, Galca AC, Aldica G, Ferreira JMF. Strong bonding between sputtered bioglass–ceramic films and Ti-substrate implants induced by atomic inter-diffusion post-deposition heat-treatments. *Appl Surf Sci* 2013;280:530–538.
12. Carlisle EM. Silicon as an essential trace element in animal nutrition. *Ciba Foundation Symp* 1986;121:123–139.
13. Hoppe A, Güldal NS, Boccaccini AR. A review of the biological response to ionic dissolution products from bioactive glasses and glass-ceramics. *Biomaterials* 2011;32(11):2757–2774.
14. Fruijtier-Pölloth C. The toxicological mode of action and the safety of synthetic amorphous silica — A nanostructured material. *Toxicology* 2012;294(2–3):61–79.
15. Vallet-Regí M, Balas F, Arcos D. Mesoporous materials for drug delivery. *Angew Chem Int Ed* 2007;46(40):7548–7558.
16. Hartmann M. Ordered mesoporous materials for bioadsorption and biocatalysis. *Chem Mater* 2005;17(18):4577–4593.
17. Vallet-Regí M, Rámila A, del Real RP, Pérez-Pariente J. A new property of MCM-41: Drug delivery system. *Chem Mater* 2001;13(2):308–311.
18. Correa-Duarte MA, Giersig M, Liz-Marzán LM. Stabilization of CdS semiconductor nanoparticles against photodegradation by a silica coating procedure. *Chem Phys Lett* 1998;286(5–6):497–501.

19. Li X, Kao F-J, Chuang C-C, He S. Enhancing fluorescence of quantum dots by silica-coated gold nanorods under one- and two-photon excitation. *Opt Exp* 2010;18(11):11335–11346.
20. Mao Z, Zhou X, Gao C. Influence of structure and properties of colloidal biomaterials on cellular uptake and cell functions. *Biomater Sci* [10.1039/C3BM00137G]. 2013;1(9):896–911.
21. Zhang Y, Hu L, Yu D, Gao C. Influence of silica particle internalization on adhesion and migration of human dermal fibroblasts. *Biomaterials* 2010;31(32):8465–8474.
22. Chithrani BD, Ghazani AA, Chan WCW. Determining the size and shape dependence of gold nanoparticle uptake into mammalian cells. *Nano Lett* 2006;6(4):662–668.
23. Yin Win K, Feng S-S. Effects of particle size and surface coating on cellular uptake of polymeric nanoparticles for oral delivery of anticancer drugs. *Biomaterials* 2005;26(15):2713–2722.
24. Costantini LM, Gilberti RM, Knecht DA. The phagocytosis and toxicity of amorphous silica. *PLoS ONE* 2011;6(2):e14647.
25. Johnston CJ, Driscoll KE, Finkelstein JN, Baggs R, O'Reilly MA, Carter J, *et al*. Pulmonary chemokine and mutagenic responses in rats after subchronic inhalation of amorphous and crystalline silica. *Toxicol Sci* 2000;56(2):405–413.
26. Chung T-H, Wu S-H, Yao M, Lu C-W, Lin Y-S, Hung Y, *et al*. The effect of surface charge on the uptake and biological function of mesoporous silica nanoparticles in 3T3-L1 cells and human mesenchymal stem cells. *Biomaterials* 2007;28(19):2959–2966.
27. Lorenz MR, Holzapfel V, Musyanovych A, Nothelfer K, Walther P, Frank H, *et al*. Uptake of functionalized, fluorescent-labeled polymeric particles in different cell lines and stem cells. *Biomaterials* 2006;27(14):2820–2828.
28. Slowing II, Wu C-W, Vivero-Escoto JL, Lin VSY. Mesoporous silica nanoparticles for reducing hemolytic activity towards mammalian red blood cells. *Small* 2009;5(1):57–62.
29. Pandurangi RS, Seehra MS, Razzaboni BL, Bolsaitis P. Surface and bulk infrared modes of crystalline and amorphous silica particles: A study of the relation of surface structure to cytotoxicity of respirable silica. *Environ Health Perspect* 1990;86:327–336.
30. Castranova V, Vallyathan V. Silicosis and coal workers' pneumoconiosis. *Environ Health Perspect* 2000;108(Suppl. 4):675–684.
31. Nabeshi H, Yoshikawa T, Matsuyama K, Nakazato Y, Matsuo K, Arimori A, *et al*. Systemic distribution, nuclear entry and cytotoxicity

of amorphous nanosilica following topical application. *Biomaterials* 2011;32(11):2713–2724.

32. Yamashita K, Yoshioka Y, Higashisaka K, Mimura K, Morishita Y, Nozaki M, *et al.* Silica and titanium dioxide nanoparticles cause pregnancy complications in mice. *Nat Nano* [10.1038/nnano.2011.41]. 2011;6(5):321–328.
33. He X, Nie H, Wang K, Tan W, Wu X, Zhang P. *In vivo* study of biodistribution and urinary excretion of surface-modified silica nanoparticles. *Anal Chem* 2008;80(24):9597–9603.
34. Huang X, Li L, Liu T, Hao N, Liu H, Chen D, *et al.* The shape effect of mesoporous silica nanoparticles on biodistribution, clearance, and biocompatibility *in vivo*. *ACS Nano* 2011;5(7):5390–5399.
35. Hench L. Biomaterials. *Science* 1980;208(4446):826–831.
36. Hench L, Wilson J. Surface-active biomaterials. *Science* 226;(4675): 630–636.
37. Hench LL, Polak JM. Third-generation biomedical materials. *Science* 2002;295(5557):1014+6-7.
38. Hench LL. Bioceramics. *J Am Ceram Soc* 1998;81(7):1705–1728.
39. Xynos ID, Edgar AJ, Buttery LDK, Hench LL, Polak JM. Ionic products of bioactive glass dissolution increase proliferation of human osteoblasts and induce insulin-like growth factor II mRNA expression and protein synthesis. *Biochem Biophys Res Commun* 2000;276(2): 461–465.
40. Xynos ID, Hukkanen MVJ, Batten JJ, Buttery LD, Hench LL, Polak JM. Bioglass® 45S5 stimulates osteoblast turnover and enhances bone formation *in vitro*: Implications and applications for bone tissue engineering. *Calcified Tissue Int* 2000;67(4):321–329.
41. Xynos ID, Edgar AJ, Buttery LDK, Hench LL, Polak JM. Gene-expression profiling of human osteoblasts following treatment with the ionic products of Bioglass® 45S5 dissolution. *J Biomed Mater Res* 2001;55(2):151–157.
42. Reffitt DM, Ogston N, Jugdaohsingh R, Cheung HFJ, Evans BAJ, Thompson RPH, *et al.* Orthosilicic acid stimulates collagen type 1 synthesis and osteoblastic differentiation in human osteoblast-like cells *in vitro*. *Bone* 2003;32(2):127–135.
43. Kawanabe K, Yamamuro T, Nakamura T, Kotani S. Effects of injecting massive amounts of bioactive ceramics in mice. *J Biomed Mater Res* 1991;25(1):117–128.

44. Kawanabe K, Yamamuro T, Kotani S, Nakamura T. Acute nephrotoxicity as an adverse effect after intraperitoneal injection of massive amounts of bioactive ceramic powders in mice and rats. *J Biomed Mater Res* 1992;26(2):209–219.
45. Lai W, Garino J, Ducheyne P. Silicon excretion from bioactive glass implanted in rabbit bone. *Biomaterials* 2002;23(1):213–217.
46. Lai W, Garino J, Flaitz C, Ducheyne P. Excretion of resorption products from bioactive glass implanted in rabbit muscle. *J Biomed Mater Res Part A* 2005;75A(2):398–407.
47. Saravanapavan P, Jones JR, Pryce RS, Hench LL. Bioactivity of gel–glass powders in the CaO–$SiO_2$ system: A comparison with ternary (CaO–$P_2P_5$–$SiO_2$) and quaternary glasses ($SiO_2$–CaO–$P_2O_5$–$Na_2O$). *J Biomed Mater Res Part A* 2003;66A(1):110–119.
48. Jones JR, Ehrenfried LM, Hench LL. Optimising bioactive glass scaffolds for bone tissue engineering. *Biomaterials* 2006;27(7):964–973.
49. Jones JR, Tsigkou O, Coates EE, Stevens MM, Polak JM, Hench LL. Extracellular matrix formation and mineralization on a phosphate-free porous bioactive glass scaffold using primary human osteoblast (HOB) cells. *Biomaterials* 2007;28(9):1653–1663.
50. Patel N, Best SM, Bonfield W, Gibson IR, Hing KA, Damien E, *et al.* A comparative study on the *in vivo* behavior of hydroxyapatite and silicon substituted hydroxyapatite granules. *J Mater Sci Mater Med* 2002;13(12):1199–1206.
51. Porter AE, Patel N, Skepper JN, Best SM, Bonfield W. Comparison of *in vivo* dissolution processes in hydroxyapatite and silicon-substituted hydroxyapatite bioceramics. *Biomaterials* 2003;24(25):4609–4620.
52. Porter AE, Patel N, Skepper JN, Best SM, Bonfield W. Effect of sintered silicate-substituted hydroxyapatite on remodelling processes at the bone–implant interface. *Biomaterials* 2004;25(16):3303–3314.
53. Botelho CM, Brooks RA, Best SM, Lopes MA, Santos JD, Rushton N, *et al.* Human osteoblast response to silicon-substituted hydroxyapatite. *J Biomed Mater Res Part A* 2006;79A(3):723–730.
54. Thian ES, Huang J, Best SM, Barber ZH, Brooks RA, Rushton N, *et al.* The response of osteoblasts to nanocrystalline silicon-substituted hydroxyapatite thin films. *Biomaterials* 2006;27(13):2692–2698.
55. Thian ES, Huang J, Best SM, Barber ZH, Bonfield W. Silicon-substituted hydroxyapatite: The next generation of bioactive coatings. *Mater Sci Eng C* 2007;27(2):251–256.

56. Aizawa M, Patel N, Porter AE, Best S, Bonfield W. Syntheses of silicon-containing apatite fibres by a homogeneous precipitation method and their characterization. *Key Eng Mater* 2006;309–311:1129–1132.
57. Kinoshita Y, Best SM, Aizawa M. *In vitro* evaluation of silicon-containing apatite fiber scaffolds for bone tissue engineering. *Key Eng Mater* 2012;529–530:391–396.
58. Nakamura J, Poologasundarampillai G, Jones JR, Kasuga T. Tracking the formation of vaterite particles containing aminopropyl-functionalized silsesquioxane and their structure for bone regenerative medicine. *J Mater Chem B* [10.1039/C3TB20589D]. 2013;1(35):4446–4454.
59. Nakamura J, Kasuga T. Enhancement of crystalline plane orientation in silsesquioxane-containing vaterite particles towards tuning of calcium ion release. *J Mater Chem B* [10.1039/C3TB21571G]. 2014; 2(9):1250–1254.
60. Obata A, Hotta T, Wakita T, Ota Y, Kasuga T. Electrospun microfiber meshes of silicon-doped vaterite/poly(lactic acid) hybrid for guided bone regeneration. *Acta Biomater* 2010;6(4):1248–1257.
61. Obata A, Tokuda S, Kasuga T. Enhanced *in vitro* cell activity on silicon-doped vaterite/poly(lactic acid) composites. *Acta Biomater* 2009;5(1):57–62.
62. Fujikura K, Lin S, Nakamura J, Obata A, Kasuga T. Preparation of electrospun fiber mats using siloxane-containing vaterite and biodegradable polymer hybrids for bone regeneration. *J Biomed Mater Res Part B* 2013;101(8):1350–1358.
63. Kasuga T, Obata A, Maeda H, Ota Y, Yao X, Oribe K. Siloxane–poly(lactic acid)–vaterite composites with 3D cotton-like structure. *J Mater Sci Mater Med* 2012;23(10):2348–2357.
64. Obata A, Ozasa H, Kasuga T, Jones J. Cotton wool-like poly(lactic acid)/vaterite composite scaffolds releasing soluble silica for bone tissue engineering. *J Mater Sci-Mater Med* 2013;24(7):1649–1658.
65. Fujikura K, Obata A, Kasuga T. Cellular migration to electrospun poly(lactic acid) fibermats. *J Biomater Sci Polym Ed* 2012;23(15): 1939–1950.

# Chapter 3

# Bioceramics for Orthopaedic Device Applications: Hydroxyapatite

**Qiaomu Tian[*,‡], Laura Rivera-Castaneda[*,§], Zachary S. Dunn[*,¶] and Huinan Liu[*,†,||]**

[*] *Department of Bioengineering, University of California, Riverside, CA 92521, USA*

[†] *Material Science and Engineering Program, University of California, Riverside, CA 92521, USA*

[‡] *qitan001@ucr.edu*
[§] *lrive012@ucr.edu*
[¶] *zdunn001@ucr.edu*
[||] *huinan.liu@ucr.edu*

## Abstract

Owing to the ageing of population, the number of orthopaedic surgeries is continuously increasing in the last few decades. Development of biocompatible, bioactive scaffolds, and implants for orthopaedic device applications attracts great research interest, as one of the most important tasks in the biomedical field. Hydroxyapatite (HAp) is a promising synthetic bioceramic for orthopaedic device applications, because it closely resembles the natural mineral of human bone. It also showed great bioactivity and biocompatibility both *in vitro* and *in vivo*. This chapter will introduce different synthesis processes of HAp and summarise different parameters that affect the resulting properties of HAp. Specific application of HAp as scaffold and as coating for orthopaedic device application will also be summarised for discussion and comparison.

## 3.1 Introduction

Human musculoskeletal system is made up of bones, muscles, tendons, ligaments, and connective tissues, which provide support and mobility to the body.[1] Musculoskeletal injuries include fractures, dislocations, sprains, strains, and traumatic amputations of neck, spine, pelvis, and extremities.[2] These conditions are the primary cause of disabilities in the United States, where the bone tissue is one of the most commonly injured types of all musculoskeletal tissues.[2] When bone tissue damages occur, reconstructive orthopaedic surgery is the preferred treatment to help the patients, including replacing the defective bone tissues with functional alternatives, or fixing and stabilising defective bone tissues with bioactive implants. Autografts and allografts are often used for these surgeries.[3,4] However, the limitations of autografts and allografts restrict them from broader applications. For example, harvesting an autograft will damage the surrounding tissue and using an allograft will cause possible immunogenic response from the patient body.[5]

Synthetic bioceramics, such as hydroxyapatite (HAp), have shown great potentials to serve as functional materials for orthopaedic device applications. As an inorganic phase, HAp accounts for 70% of dry weight of natural human cortical bones.[6] It deposits parallel to collagen fibres and provides stiffness to the bone.[6] Synthetic HAp have also shown excellent biocompatibility, bioactivity, and osteoconductivity both *in vitro* and *in vivo*.[7–9] Researchers developed HAp-based scaffolds and HAp-based coatings on metallic substrates for orthopaedic device applications for few decades. This chapter will summarise different processes of synthesising HAp particles, biological properties of

HAp, as well as current and potential applications of HAp in orthopaedic devices.

## 3.2 Synthesis of HAp Particles

Both in laboratory and industry, HAp particles can be produced in different processes, as indicated in the tree diagram of Figure 3.1. The properties of resulting HAp particles, such as morphology, size, crystallinity, are greatly influenced by the applying parameters, which will also be discussed along each process.

### 3.2.1 Wet chemistry precipitation of HAp

Wet chemistry precipitation followed by hydrothermal treatment is a widely used method to produce HAp particles.[10,11] This method involves the addition of phosphate ion solution to a calcium ion

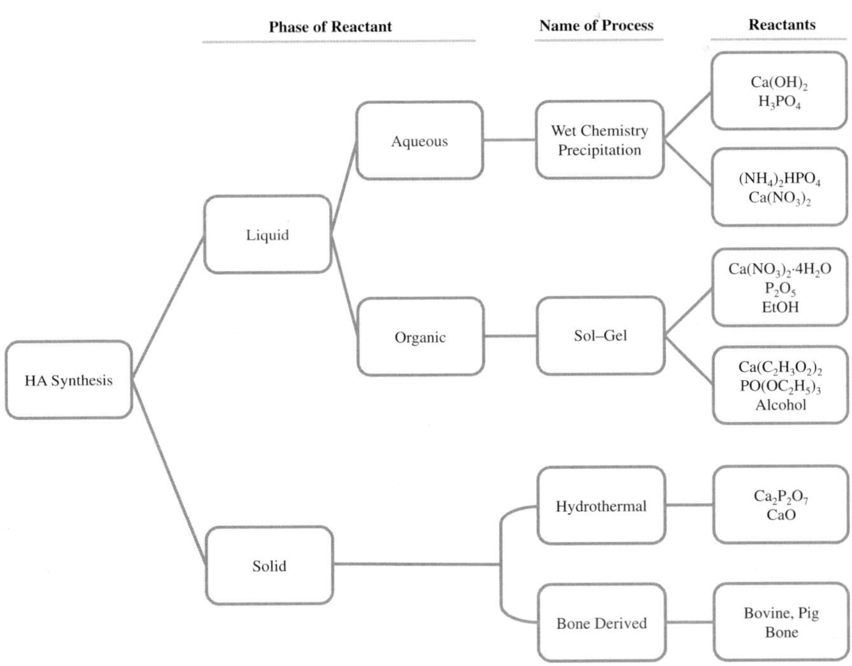


Figure 3.1 Tree diagram for different synthesis processes of HAp particles.

solution to precipitate HAp. Various reactions can be applied to obtain HAp through wet chemistry precipitation. One common reaction to be followed is

$$10Ca(NO_3)_2 + 6(NH_4)_2HPO_4 + 8NH_4OH \\ \rightarrow Ca_{10}(PO_4)_6(OH)_2 + 6H_2O + 20NH_4NO_3 \quad (3.1)$$

This reaction adds a calcium nitrate ($Ca(NO_3)_2$) solution dropwise to a diammonium hydrogen phosphate ($(NH_4)_2HPO_4$) solution at a constant temperature.[12] Ammonium hydroxide ($NH_4OH$) is used to maintain the pH above 10 during the reaction. The addition of the calcium nitrate yields the precipitation of a calcium phosphate, which is subsequently centrifuged, washed several times, resuspended, and then hydrothermally treated. After the hydrothermal treatment, HAp is collected and dried under vacuum, ground with a mortar, and pestle to obtain a fine powder.[12]

In addition, HAp synthesis through wet chemistry precipitation can also be carried out with a different set of reactants as follows:

$$5Ca(OH)_2 + 3H_3PO_4 \rightarrow Ca_5(PO_4)_3(OH) + 9H_2O \quad (3.2)$$

Similar to the previous approach, this reaction entails the dropwise addition of orthophosphoric acid ($H_3PO_4$) to a suspension of calcium hydroxide ($Ca(OH)_2$) to cause the precipitation of HAp at a constant temperature and pH. To obtain the final HAp product, the precipitates should be washed several times to remove any remaining soluble salts and then dried to evaporate remained water. The precipitates are collected and either crushed, ground, or sieved to obtain desired particles.

For wet chemistry precipitation approaches, varying reaction parameters, such as temperature and reaction time, can affect the properties of the final HAp particles. For example, when following the first chemical reaction (Eq. (3.1)), increasing the reaction time at room temperature resulted in size reducing of the HAp particles.[13] Particle size of HAp can also be influenced by the temperature, where raising it leads to larger HAp particles.[14] On the other hand, raising the temperature can also lead to a needle-shaped morphology, whereas lowering it causes the particle morphology to become more spherical.[15] When following the second chemical reaction (Eq. (3.2)), raising the temperature can lead to a spherical morphology rather needle-shaped morphology.[15] Additionally, the morphology may also be influenced by the rate of orthophosphoric acid addition, and a decreased addition rate leads to needle-shaped particles.[16]

### 3.2.2 Sol–gel synthesis of HAp

Sol–gel synthesis of HAp is a wet chemistry method that usually involves the dissolution and mixing of calcium and phosphorus precursors in an organic solvent. A gel should form in the mixture during constant heating. A drying and sintering stage should follow and the resulting product should be crushed to obtain HAp fine powder.[17,18] The sol–gel technique is advantageous because it involves molecular-level mixing of calcium and phosphorus precursors, thus providing a more homogeneous HAp product. Additionally, the high reactivity of the sol–gel powders allows for reactions under low temperature.[18] Although it has been reported that following the sol–gel method yields highly pure product, there have also been cases of a secondary phase of calcium oxide being present, most notably when phosphorous alkoxide was used as a phosphorous precursor. Another widely known limitation of the sol–gel method has been the series of complicated steps that must followed to obtain pure HAp after heat treatment.[19]

### 3.2.3 Hydrothermal synthesis of HAp

Hydrothermal synthesis involves a reaction of calcium and phosphorus precursors under an aqueous environment at high temperature and pressure within a sealed system, usually ranging from 80 °C to 400 °C and autogenous pressure to 100 MPa, respectively.[20] HAp particles can be synthesised with hydrothermal treatment of beta-tricalcium phosphate ($\beta$-TCP) and citric acid, or calcium nitrate, diammonium hydrogen phosphate, and urea.[21,22] HAp nanoparticles have also been synthesised through the use of three-pump hydrothermal flow system with calcium nitrate and ammonium hydrogen phosphate solutions. This method required only one step and lasted a few seconds rather than hours, as is the case for those previously described.[23]

The morphologies of HAp particles produced by hydrothermal synthesis varies depending on the applied parameters in the process. For example, hexagonal HAp particles can be obtained by adding chelate, such as ethylene diamine tetra acetic acid, which placed itself on the surface of HAp, and allowed for a closer resemblance to a hexagonal prism as the temperature increased.[24] In addition, pH and temperature of the reaction can also affect morphology of HAp particles.[25] It has been shown that pH of 6 and 9 resulted in whisker-like morphologies, while pH of 14 did not. At pH of 14, HAp particles had lower crystallinity when compared to that of HAp particles which produced at pH of 6 and 9. Decreasing temperature allowed for the formation of HAp whisker-like nanoparticles with higher aspect ratio.[25]

### 3.2.4 Bone-derived HAp

Other than synthesizing HAp by chemical reactions, HAp can be derived from animal bones to allow for the retention of bone properties, such as chemical composition. The process begins with obtaining fresh animal (e.g., bovine, pig, etc.) bones and removing any soft tissues still attached. Subsequent thermal treatment is used to remove internal contents, water, and also ensure that all disease-causing agents are eliminated. Afterwards, remaining materials are crushed to a fine HAp powder.[26,27] It has been suggested that bone-derived HAp have better osteoconductivity and osteoinductivity than synthetic HAp because it can retain the composition.[27] The bone-derived process may also have the advantage of allowing high yield, in contrast to other synthetic processes, which usually yield small quantities of HAp.[26]

Representative scanning electron microscopic images of HAp particles produced by wet chemistry precipitation, sol–gel, hydrothermal, and bone-derived were summarised in Figure 3.2.

## 3.3 Properties of HAp Particles

### 3.3.1 Biological properties and functions of HAp particles

Osteoconductivity is the ability of bone cells to attach and grow on the surface or grow into the pores/channels of a biomaterial, which is a desired property for orthopaedic implants.[28] Adsorption of protein on the surface of a material is the initial step of cell–biomaterial interaction. Proteins, such as albumin, can be adsorbed onto

Figure 3.2 Representative scanning electron microscopic images of HAp particles produced by (a) wet chemistry precipitation, (b) sol–gel, (c) hydrothermal, and (d) bone-derived. Images adapted from Refs. 19, 21, 27, and 105, respectively.

the surface of HAp particles.[29,30] In addition, HAp particles showed enhancement in adsorption of fibronectin and vitronectin on the composites scaffold.[31] Secondly, the presence of HAp can facilitate osteoblast adhesion and proliferation *in vitro*, which is an indication of good osteoconductivity.[9,32] Moreover, HAp particles have also shown osteoinductivity, which is the ability of a material to induce stem cells to differentiate and develop into mature bone cells.[28] Liu *et al.*[33] indicated that nanophase HAp could promote the differentiation of mesenchymal stem cells (MSCs) into osteogenic cells.

### 3.3.2 Size effect

By adjusting synthesis parameters, nano- and micron-size HAp particles can be produced. Significant differences were found between different sizes of HAp particles. Guo *et al.*[34] found that HAp nanoparticles were more efficient in the adhesion and proliferation of human primary osteoblast cells than micron particles. Additionally, HAp nanoparticles improved mineralisation more than HAp micron particles, which is important for bone regeneration.[34] These findings paralleled those by Webster *et al.*[9] where it was found that a ceramic's grain size was inversely related to the adhesion of osteoblasts. These results may attribute to the fact that nanosized HAp resemble the natural mineral in bone more closely than its micron-size counterpart.[35]

Significant differences have also been found within the nanoscale level as well. Cai *et al.* synthesised HAp nanoparticles of three different diameters (20 ± 5 nm, 40 ± 10 nm, and 80 ± 12 nm) and conducted an *in vitro* study using MSCs and osteosarcoma cells (U2OS). It was found that a smaller HAp nanoparticle size increased proliferation of MSCs but decreased proliferation of U2OS cells.[35] Shi *et al.*[32] also indicated that smaller nanosize of HAp nanoparticles increased the proliferation of human osteoblast-like cells. It was speculated that decreasing HAp nanoparticle size will increase their penetration of the cell membrane and probability of inducing a response from the cell, i.e., proliferation and growth.[32,35]

### 3.3.3 Morphology effect

Different morphologies of HAp, which were summarised in Figure 3.3, can also affect its property significantly. Zhao *et al.*[36] indicated that nanophase HAp with spherical and rod shape resulted in significantly less cell death after culture with human bronchial epithelial cell, compared to that of nanophase HAp with needle and plate shape. Conversely, no significant cell death was recorded in mouse macrophage cell culture when exposed to any of the nanophase HAp groups.[36] Furthermore, Shi *et al.* examined the effect of two different types of morphologies (rod- and sphere-like) on human osteoblast-like (MG-63) cells. Sphere-like HAp

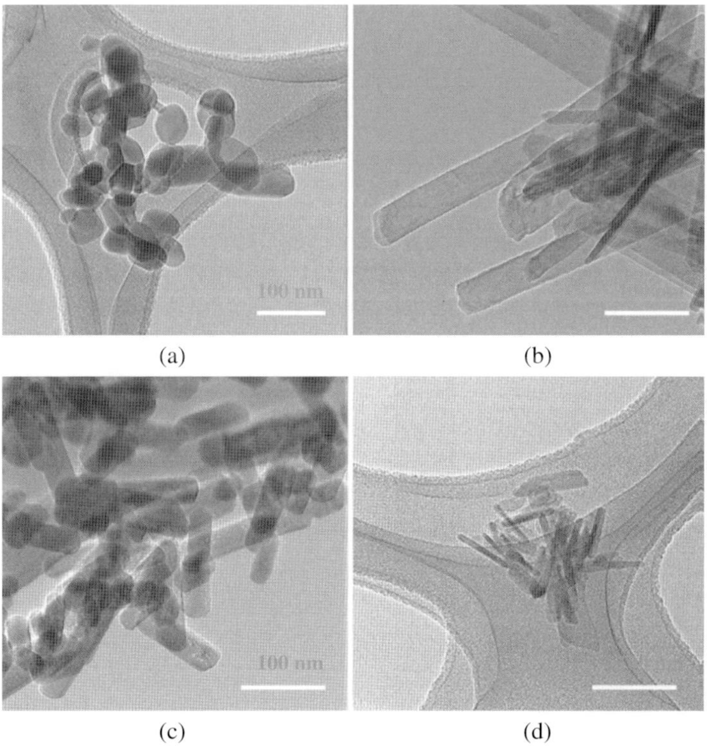


Figure 3.3 Representative transmission electron microscopic images of HAp particles with (a) sphere, (b) plate, (c) rod, and (d) needle shape. Scale bar = 100 nm. Images adapted from Ref. 36.

nanoparticles induced higher proliferation of MG-63 cells than rod-like HAp nanoparticles.[32] Similarly, Zhao *et al.* showed that a spherical shape of HAp nanoparticles caused greater adhesion and proliferation of osteoblasts than rod-like nanoparticles. This may be due to the fact that sphere-like nanoparticles had higher surface area than the rod-like nanoparticles.[37]

### 3.3.4 Metal ions substitution

The biological property of HAp particles can also be improved by substitution of metal ions, such as magnesium (Mg), strontium (Sr), cobalt (Co), and zinc.[38,39] Wet chemistry precipitation and sol–gel process are adopted to dope different metal ions within HAp during the synthesis process.[39] Landi *et al.*[40] indicated that Mg-substituted HAp had greater osteoconductivity and higher material resorption than the commercial pure HAp *in vivo*. In addition, Kim *et al.*[41] prepared Sr-substituted HAp using precipitation method and indicated that the Sr-substituted HAp had an improvement in hardness, as well as a better proliferation of osteoblast-like cells compared with pure HAp. Multiple ions substitution of HAp has also achieved by researchers. For example, Mg and Co ions have been doped within HAp by Kulanthaivel *et al.*[42] and they indicated that dual doping HAp showed significantly higher protein absorption capacity in comparison to pure HAp. In addition, the doping significantly promoted cell proliferation and differentiation of osteoblast cells, as well as a higher expression of vascular endothelial growth factor, a key molecule that promotes angiogenesis.[42]

## 3.4 Applications

### 3.4.1 HAp-based scaffolds

Based on the aforementioned attracting properties of HAp, three-dimensional (3D) HAp-based scaffolds are promising to promote bone growth through interactions with bone cells and to guide new bone tissue regeneration as orthopaedic implants.[43] In general, these scaffolds are designed with porosity to resemble human bones, which permit the fluid exchange and promote the adhesion, growth, and differentiation of cells. Porous HAp single-phase scaffolds can be synthesised through thermal treatment, freeze-drying, compacting, and sintering.[44,45] HAp-based composite scaffolds could be made through different synthesis process, such as porogen leaching, phase separation, and gas foaming.[43,46–48] In addition, 3D printing and rapid prototyping have also been adopted for fabricating HAp-based scaffolds in recent years, which are most promising to achieve individualised implant design for patients.[49–51]

#### 3.4.1.1 *HAp single-phase scaffold*

HAp single-phase scaffold has been widely investigated for biomedical applications. For example, Pro Osteon is a commercially available porous HAp scaffold from Biomet, Inc. This HAp scaffold is produced by hydrothermal treatment of marine corals with ammonium hydrogen phosphate and water.[52] The reaction involves the conversion of the calcium carbonate porous structure of the coral sample to HAp under high temperature and pressure, which sterilises the HAp while maintaining its original porous structure and nanosized HAp crystals.[53–55] Since the porous structure of the original coral is preserved, scaffolds can be used for orthopaedic device applications such as bone grafting.[55]

Porous HAp scaffolds can also be produced by compacting and sintering the HAp particles into a single solid mass. The porosity of HAp scaffolds can be controlled by inserting polymeric rods during compacting, which will then evaporate during subsequent sintering. Oliveira *et al.*[56] found bone marrow stromal cells could not only adhere and proliferate on the surface of high-porosity HAp scaffolds, but also grow into the scaffold *in vitro*. In addition, Tamai *et al.*[57] created HAp scaffolds with random interconnectivity of spherical pores and implanted them into the femoral condyles of rabbits. It has been showed that bone and bone marrow with vessels formed deeply in the scaffold through the porous structure.[57] Additionally, the compressive strength of the HAp scaffold increased over time, which showed that bone growth reinforced the overall structure.[57]

#### 3.4.1.2 *HAp composite scaffold*

HAp composite scaffolds are also studied and produced by supplemented HAp with other materials, such as polymers, metals, and oxides.[58] Biocompatible polymers often provide additional favourable properties for HAp scaffold as reinforced materials, where they allow for enhanced mechanical properties and alleviate the brittleness of HAp.[46] In addition, a HAp/polymer composite scaffold would also retain the bioactivity of the HAp to promote bone growth rather than the relative bioinert properties of single-phase polymers.[47]

Poly(lactic-co-glycolic) acid (PLGA) is a biocompatible and biodegradable polymer and has been paired with HAp in composite scaffold. Kim *et al.*[46] indicated that HAp in the HAp/PLGA composite scaffold improved bone tissue regeneration during an *in vivo* study of rats. Aside from better bone growth, it was found that HAp/PLGA composite scaffold also showed improvement in compressive modulus and tensile modulus. Other conducted *in vivo* studies also indicated that HAp/PLGA composites with seeded osteogenic cells enhanced new bone tissue formation when implanted in mice.[59] Polycaprolactone (PCL) is another polymer often in conjunction with HAp for composite scaffold application.[60] PCL is a biocompatible and biodegradable material, which will degrade in several years *in vivo*.[61,62] Zhao *et al.*[63] indicated that HAp/PCL composite scaffold had improved compressive strength than a single-phase HAp scaffold. Heo *et al.*[64] found that osteoblast-like cells attached, grew, and proliferated on HAp/PCL composite scaffold.

Polyetheretherketone has been noted as a non-degradable polymer that can potentially be used for load-bearing applications because of its attractive mechanical properties.[65–67] This polymer can also be applied with HAp to form a composite scaffold, where the Young's modulus and tensile strength of HAp/PEEK composite scaffold was found to be similar to those of cortical bone.[68] Also, an *in vivo* study showed the presence of fibroblast cells and formation of bone within the pores of the HAp/PEEK scaffold. To further improve the mechanical property, HAp/PEEK scaffolds could also reinforce with strontium (Sr), which showed to promote the formation of bone and prevent new bone resorption.[69] It was found that the Sr-HAp/PEEK scaffold had a controllable elastic modulus by changing the Sr-HAp content. The addition of Sr was proved to increase the proliferation of osteoblasts better than HAp on its own.[70]

### 3.4.2 HAp-coated implants for load-bearing applications

Metallic materials, e.g., titanium alloys and stainless steels, are often used to immobilise bone fractures because they have the required mechanical strength. HAp coatings on such metallic devices can provide a bioactive interface between the metallic implant and the surrounding human tissues and also improve the antibacterial property, osteointegration, and biocompatibility of the implant.[71–73]

#### 3.4.2.1 *Plasma spray*

A variety of HAp coating techniques have been investigated. One of the most popular and commercialised techniques for HAp coating is plasma spray. Plasma spray is a type of thermal spraying technique, where plasma is used to heat up HAp to high temperature and causes it to move at high velocities and hit the substrate, creating a coating.[74,75] Representative literatures reporting plasma-sprayed single-phase HAp and HAp composite coating on metallic substrates were summarised and listed in Table 3.1.

Maintaining the pure crystalline phase of HAp is important for a coating because it closely resembles the natural bone and therefore enhances the bioactivity of bone–implant interface.[73] Under high temperatures of plasma spray, however, it is usually hard to control the microstructure and stoichiometry of HAp which resulted in secondary crystalline phases. In addition, amorphous phase may form at the interface between the coating

Table 3.1 Literature reviews of HA-based coatings on metallic substrates produced by plasma spray.

| Coating material(s) | Substrate material(s) | Mechanical strength | *In vitro* | *In vivo* | References |
|---|---|---|---|---|---|
| HAp | Titanium alloy (Ti–6Al–4V) | Shear strengths: $4.31 \pm 2.73$ and $10.59 \pm 1.04$ MPa<br>Bonding strengths: $27.15 \pm 1.75$ and $50.21 \pm 2.14$ MPa | N/A | Lateral cortices of mongrel dogs for 4, 6, 8, and 12 weeks | [74] |
| HAp/titanium | Titanium alloy (Ti–6Al–4V) | Bonding strengths: 14.5 and 17.3 MPa | Immersion in SBF for 14 days | N/A | [75] |
| HAp/bioactive glass | Titanium alloy (Ti–6Al–4V) | Bonding strengths: 50–60 MPa | Immersion for 30 days | N/A | [80] |
| HAp/CNTs | Titanium | Fracture toughness: $0.61 \pm 0.09$ MPa $m^{1/2}$ | Induced human osteoblast cell for 3 days | N/A | [77] |
| HAp/strontium | Titanium alloy (Ti–6Al–4V) | Bonding strength: 23.5 MPa | Immersion in SBF for 1 and 3 days; Clones osteoblastic precursor cell line 1 for 3, 10, and 28 days | N/A | [78] |

and the substrate due to high temperature as well. Although the amorphous phase may promote bone growth, its fast resorption rate compromise fixation and cause the release of debris to the surrounding tissues.[75]

Composite coatings are proposed to potentially address the limitations of plasma-sprayed single-phase HAp coating while retaining its bioactivity and biocompatibility. For example, HAp/titanium (Ti) composite coatings have been deposited on a titanium alloy substrate, which showed decrease in the presence of the amorphous phase as well as increase in the bonding strength of the coating.[76] Carbon nanotubes (CNTs) can also be used to create composite coating on metallic substrates. The addition of CNTs enhanced the crystallinity of HAp and fracture toughness of the coatings, as well as promoted osteoblast cell growth when compared to single-phase HAp coating.[77] Other attempts, such as adding bioactive glass (BG), Sr and Ti oxide, were also carried out by the researchers.[78,79] It was found that, however, as more BG was included, the HAp start to decompose to other phases, such as $\alpha$-TCP and $\beta$-TCP.[80] Although various composite coatings have been achieved, it is unlikely to incorporate polymers and bioentity such as antibody, proteins, and drugs which have low tolerance to heat in the HAp coating using plasma spray.

### 3.4.2.2 *Electrodeposition*

Electrodeposition can be carried at low temperatures and on irregularly shaped substrates, where the low processing temperature prevents the formation of secondary phases.[81] It involves the use of an aqueous electrolytic solution containing necessary ingredients to obtain HAp, such as calcium nitrate ($Ca(NO_3)_2$) and ammonium hydrogen phosphate ($NH_4H_2PO_4$).[82] The substrate is connected to a power source and a current is applied, allowing for the chemical reaction and deposition to occur on the substrate, which is usually a cathode.[83,84] Representative literatures reported on single-phase HAp and HAp composite coating produced by electrodeposition were summarised in Table 3.2.

Table 3.2 Literature reviews of HA-based coatings on metallic substrates produced by electrodeposition.

| Coating material(s) | Substrate material(s) | Mechanical strength | *In vitro* | *In vivo* | References |
|---|---|---|---|---|---|
| HAp | Titanium alloy (Ti–6Al–4V) | Low bonding strength | N/A | Canine trabecular bone for 6 h, 7 days, and 14 days | [85] |
| | Magnesium alloy (AZ91D); pure magnesium | N/A | Immersion in SBF for 48 h | N/A | [82, 84] |
| | Stainless Steel (316L) | N/A | Immersion in SBF for 0, 1, 3, 5, 8, 10, 14, and 17 days | N/A | [83] |
| HAp/CNTs | Pure titanium | Adhesion strength: 24.2 ± 0.8 MPa<br>Hardness: 6.72 ± 0.45 GPa<br>Elastic modulus: 125.7 ± 4.0 MPa | L929 mouse fibroblast cells for 48 h | N/A | [87] |
| HAp/calcium silicate | Titanium | Bonding strength: 19.1 ± 4.7 MPa | Mouse osteoblast cells for 1 and 7 days | N/A | [86] |
| HAp/manganese | Titanium | Bonding strength: 13.9 ± 3.0 MPa | Mouse osteoblast cells for 3 and 7 days | N/A | [85] |

Electrodeposition was used to modify the surface of Ti substrates with HAp coating. Bone-like tissue appeared on both electrodeposited and plasma-sprayed HAp coatings on Ti–6Al–4V substrates after 14 days of implantation in a canine trabecular bone.[85] The morphology of the electrodeposited HAp was favourable in the integration of new bone tissue, but the bonding strength between was still low.[85] Electrodeposition can also produce composite coatings. For example, Huang *et al.*[86] deposited HAp coating supplemented with calcium silicate ($CaSiO_3$) or manganese (Mn) on Ti substrate. Both the HAp/$CaSiO_3$ and HAp/Mn composite coatings were found to have superior bonding strength and resistance to corrosion than a HAp coating.[86] Composite coating consisting of HAp and CNTs can also be applied to a Ti substrate, where the composite coating had enhanced resistance to corrosion and bonding strength in comparison to a HAp coating. Following an *in vitro* test using mouse fibroblasts, the HAp/CNTs composite coating also showed better cell viability than a comparable HAp coating.[87]

#### 3.4.2.3 *Electrophoretic deposition*

Electrophoretic deposition (EPD) is implemented utilising a suspension containing charged particles, which are deposited on an oppositely charged substrate when voltage is applied.[88,89] It has been noted that water is not a favourable liquid medium due to its gas-releasing reactions under voltage, as compared to organic solvents.[90,91] EPD allows for the coating of substrates with a variety types, such as Ti, stainless steel (SS), nitinol, and magnesium (Mg). EPD can also be applied on substrates with different shapes, such as plain, foil, wire, and cylindrical. Representative literatures reporting on EPD-coated single-phase HAp and HAp composite coating were summarised in Table 3.3.

For EPD, altering the voltage applied and the deposition time allows for the control of resulting coating thickness and morphology.[90,91] Javidi *et al.*[92] indicated that the thickness of HAp

Table 3.3 Literature reviews of HA-based coatings on metallic substrates produced by EPD.

| Coating material(s) | Substrate material(s) | Mechanical strength | *In vitro* | *In vivo* | References |
|---|---|---|---|---|---|
| HAp | Titanium alloy (Ti–6Al–4V) | Hardness: 683 Hv<br>Bonding strength: 8.5 MPa | Immersion in Hanks' solution for 4 weeks | N/A | [89] |
| | | Bonding strength: 18 MPa | Immersion in SBF for 6 weeks | Calf vertebra for 12 weeks | [93] |
| | Magnesium alloy (AZ91) | N/A | Immersion in SBF for 28 days | N/A | [90] |
| HAp/CNTs | Pure titanium | N/A | Mouse osteoblast cells for 2 and 4 days | N/A | [96] |
| HAp/chitosan | Stainless steel (316L) | Bonding strength: 0.1, 1.35, and 0.95 MPa | N/A | N/A | [98] |
| HAp/titania | Titanium | Bonding strength: 48 ± 3.67 MPa | Human osteosarcoma cell line for 48 h | N/A | [95] |
| HAp/zirconia | Titanium alloy (Ti–6Al–7Nb) | Removal torque: 62 N cm | N/A | Adult New Zealand white rabbit for 2, 6, and 18 weeks | [97] |

coating on SS increased by increasing the voltage applied as well as the deposition time, while optimal parameters could produce coatings with no apparent cracks. Aksakal[93] indicated that HAp-coated Ti screws showed increase in calcium concentration around implantation sites. The HAp coating produced by EPD normally have no sufficient bonding strength to the substrates.[94] Sintering after EPD can improve the adhesion strength of the coating, however it may affect the crystallinity of HAp.[94] Composite coatings are also proposed to potentially improve the adhesion and reinforce the mechanical properties of HAp coating produced by EPD. CNTs have been paired with HAp to improve the adhesion, corrosion resistance, and biocompatibility as composite coating.[95,96] Alzubaydi *et al.*[97] indicated that HAp/zirconia composite coating showed high cellular activity and mature bone tissue regeneration *in vivo*.

EPD is one of the most promising techniques to conduct complex biocomposite coating, including HAp with polymers such as PLGA, chitosan, and alginate. The low processing temperature and charge-derived mechanism can maximally maintain the microstructure, stoichiometry, and chemistry of the material after coating. Studies have been extensively carried out for HAp/polymer composite coatings on metallic substrates by EPD.[98–104] For example, HAp/chitosan composite coating on SS by EPD indicated an improvement of corrosion resistance and sufficient bioactivity.[102] HAp/PLGA composites on Mg substrates by EPD showed improved corrosion resistance of the Mg substrates as well.[105] Another possible incorporation in the coating is bioentity, such as growth factors, proteins, drugs, and antibodies, which is also promising to achieve by EPD, although to the author's knowledge, no research has reported so far.

Representative scanning electron microscopic images of single-phase HAp coating and HAp composite coating produced by plasma spray, electrodeposition, and EPD were summarised in Figure 3.4.

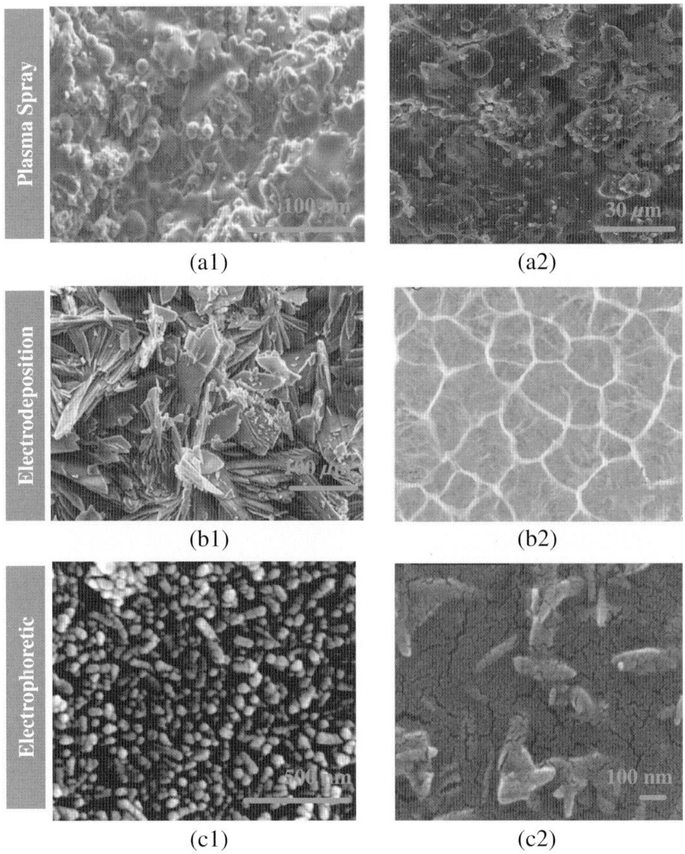


Figure 3.4 Representative scanning electron microscopic images of (a) plasma spray, (b) electrodeposition, and (c) electrophoretic deposition produced (1) single-phase HAp and (2) HAp composite coating. Images adapted from Refs. 78, 79, 84, 86, 102, and 105, respectively.

## 3.5 Conclusion

Research and development of functional scaffolds and implants are projected to be continuously growing in the biomedical field. HAp will also be an important research topic as a biocompatible and bioactive ceramic for orthopaedic device applications. Wet chemistry precipitation can produce highly crystallised HAp but with limited quantity. For industry application, larger production such as bone-derived HAp still needs optimisation to control the crystallinity of HAp. Biocomposite scaffold and coatings are more promising than the single-phase HAp in terms of tunable mechanical properties and improved bioactivity. Research efforts are still projected to develop a stable and tunable method for producing biocomposite scaffold or coating that can incorporate various materials and bioentities and also achieving individualised design of implants.

## References


1. Liu HN. (ed.) *Biodegradable metals and responsive biosensors for musculoskeletal Applications*, Nanotechnology Enabled *in Situ* Sensors for Monitoring Health. New York: Springer; 2011;115–137.
2. Surgeons AAoO. *The Burden of Musculoskeletal Diseases in the United States: Prevalence, Societal and Economic Cost*. Rosemont, IL; 2008.
3. Kalfas IH. Principles of bone healing. *Neurosurg Focus* 2001;10:E1.
4. Palmer RH, Hulse DA, Hyman WA, Palmer DR. Principles of bone healing and biomechanics of external skeletal fixation. *Vet Clin North Am Small Anim Pract* 1992;22:45–68.
5. Betz RR. Limitations of autograft and allograft: New synthetic solutions. *Orthopedics* 2002;25:S561.
6. Webster TJ, Ahn ES. Nanostructured biomaterials for tissue engineering bone. *Adv Biochem Eng/Biotechnol* 2007;103:275–308.
7. Liu HN, Webster TJ. Ceramic/polymer nanocomposites with tunable drug delivery capability at specific disease sites. *J Biomed Mater Res Part A* 2010;93A:1180–1192.
8. Liu HN, Yazici H, Ergun C, Webster TJ, Bermek H. An *in vitro* evaluation of the Ca/P ratio for the cytocompatibility of nano-to-micron particulate calcium phosphates for bone regeneration. *Acta Biomater* 2008;4:1472–1479.
9. Webster TJ, Siegel RW, Bizios R. Osteoblast adhesion on nanophase ceramics. *Biomaterials* 1999;20:1221–1227.
10. Correia RN, Magalhaes MCF, Marques PAAP, Senos AMR. Wet synthesis and characterization of modified hydroxyapatite powders. *J Mater Sci-Mater M* 1996;7:501–505.
11. Tian QM, Liu HN. Electrophoretic deposition and characterization of biocomposites on magnesium for orthopedic applications. *Adv Mater Res* 2014;922:761–766.

12. Johnson I, Akari K, Liu H. Nanostructured hydroxyapatite/poly(lactic-co-glycolic acid) composite coating for controlling magnesium degradation in simulated body fluid. *Nanotechnology* 2013;24:375103.
13. Webster TJ, Ergun C, Doremus RH, Siegel RW, Bizios R. Enhanced functions of osteoblasts on nanophase ceramics. *Biomaterials* 2000;21:1803–1810.
14. Mobasherpour I, Heshajin MS, Kazemzadeh A, Zakeri M. Synthesis of nanocrystalline hydroxyapatite by using precipitation method. *J Alloys Compd* 2007;430:330–333.
15. Kumar R, Prakash KH, Cheang P, Khor KA. Temperature driven morphological changes of chemically precipitated hydroxyapatite nanoparticles. *Langmuir* 2004;20:5196–5200.
16. Bouyer E, Gitzhofer F, Boulos MI. Morphological study of hydroxyapatite nanocrystal suspension. *J Mater Sci-Mater M* 2000;11:523–531.
17. Jillavenkatesa A, Condrate RA. Sol–gel processing of hydroxyapatite. *J Mater Sci* 1998;33:4111–4119.
18. Feng W, Mu-Sen L, Yu-Peng L, Yong-Xin Q. A simple sol–gel technique for preparing hydroxyapatite nanopowders. *Mater Lett* 2005;59:916–919.
19. Kim I-S, Kumta PN. Sol–gel synthesis and characterization of nanostructured hydroxyapatite powder. *Mater Sci Eng B* 2004;111:232–236.
20. Zakaria SM, Sharif Zein SH, Othman MR, Yang F, Jansen JA. Nanophase hydroxyapatite as a biomaterial in advanced hard tissue engineering: a review. *Tissue Eng Part B Rev* 2013;19:431–441.
21. Yoshimura M, Suda H, Okamoto K, Ioku K. Hydrothermal synthesis of biocompatible whiskers. *J Mater Sci* 1994;29:3399–3402.
22. Neira IS, Guitián F, Taniguchi T, Watanabe T, Yoshimura M. Hydrothermal synthesis of hydroxyapatite whiskers with sharp faceted hexagonal morphology. *J Mater Sci* 2007;43:2171–2178.
23. Chaudhry AA, Haque S, Kellici S, Boldrin P, Rehman I, Khalid FA, Darr JA. Instant nano-hydroxyapatite: A continuous and rapid hydrothermal synthesis. *Chem Commun* 2006;21:2286–2288.
24. Zhu R, Yu R, Yao J, Wang D, Ke J. Morphology control of hydroxyapatite through hydrothermal process. *J Alloys Compd* 2008;457:555–559.
25. Liu JB, Ye XY, Wang H, Zhu MK, Wang B, Yan H. The influence of pH and temperature on the morphology of hydroxyapatite synthesized by hydrothermal method. *Ceram Int* 2003;29:629–633.

26. Ruksudjarit A, Pengpat K, Rujijanagul G, Tunkasiri T. Synthesis and characterization of nanocrystalline hydroxyapatite from natural bovine bone. *Curr Appl Phys* 2008;8:270–272.
27. Doostmohammadi A, Monshi A, Fathi MH, Braissant O. A comparative physico-chemical study of bioactive glass and bone-derived hydroxyapatite. *Ceram Int* 2011;37:1601–1607.
28. Albrektsson T, Johansson C. Osteoinduction, osteoconduction and osseointegration. *Eur Spine J* 2001;10(Suppl. 2):S96–S101.
29. Mura-Galelli MJ, Voegel JC, Behr S, Bres EF, Schaaf P. Adsorption/desorption of human serum albumin on hydroxyapatite: A critical analysis of the Langmuir model. *Proc Natl Acad Sci USA* 1991;88: 5557–5561.
30. Wassell DTH, Hall RC, Embery G. Adsorption of bovine serum-albumin onto hydroxyapatite. *Biomaterials* 1995;16:697–702.
31. Woo KM, Seo J, Zhang R, Ma PX. Suppression of apoptosis by enhanced protein adsorption on polymer/hydroxyapatite composite scaffolds. *Biomaterials* 2007;28:2622–2630.
32. Shi Z, Huang X, Cai Y, Tang R, Yang D. Size effect of hydroxyapatite nanoparticles on proliferation and apoptosis of osteoblast-like cells. *Acta Biomater* 2009;5:338–345.
33. Liu Y, Wang G, Cai Y, Ji H, Zhou G, Zhao X, Tang R, Zhang M. *In vitro* effects of nanophase hydroxyapatite particles on proliferation and osteogenic differentiation of bone marrow-derived mesenchymal stem cells. *J Biomed Mater Res Part A* 2009;90:1083–1091.
34. Guo X, Gough JE, Xiao P, Liu J, Shen Z. Fabrication of nanostructured hydroxyapatite and analysis of human osteoblastic cellular response. *J Biomed Mater Res Part A* 2007;82:1022–1032.
35. Cai Y, Liu Y, Yan W, Hu Q, Tao J, Zhang M, Shi Z, Tang R. Role of hydroxyapatite nanoparticle size in bone cell proliferation. *J Mater Chem* 2007;17:3780.
36. Zhao X, Ng S, Heng BC, Guo J, Ma L, Tan TT, Ng KW, Loo SC. Cytotoxicity of hydroxyapatite nanoparticles is shape and cell dependent. *Arch Toxicol* 2013;87:1037–1052.
37. Zhao Y, Zhang Y, Ning F, Guo D, Xu Z. Synthesis and cellular biocompatibility of two kinds of HAP with different nanocrystal morphology. *J Biomed Mater Res Part B* 2007;83:121–126.
38. Haberko K, Bućko MM, Brzezińska-Miecznik J, Haberko M, Mozgawa W, Panz T, Pyda A, Zarębski J. Natural hydroxyapatite — Its behaviour during heat treatment. *J Eur Ceram Soc* 2006;26:537–542.

39. Šupová M. Substituted hydroxyapatites for biomedical applications: A review. *Ceram Int* 2015;41:9203–9231.
40. Landi E, Logroscino G, Proietti L, Tampieri A, Sandri M, Sprio S. Biomimetic Mg-substituted hydroxyapatite: From synthesis to *in vivo* behaviour. *J Mater Sci-Mater M* 2008;19:239–247.
41. Kim HW, Koh YH, Kong YM, Kang JG, Kim HE. Strontium substituted calcium phosphate biphasic ceramics obtained by a powder precipitation method. *J Mater Sci Mater M* 2004;15:1129–1134.
42. Kulanthaivel S, Mishra U, Agarwal T, Giri S, Pal K, Pramanik K, Banerjee I. Improving the osteogenic and angiogenic properties of synthetic hydroxyapatite by dual doping of bivalent cobalt and magnesium ion. *Ceram Int* 2015;41:11323–11333.
43. Wei G, Ma PX. Structure and properties of nano-hydroxyapatite/polymer composite scaffolds for bone tissue engineering. *Biomaterials* 2004;25:4749–4757.
44. Fu Q, Rahaman MN, Dogan F, Bal BS. Freeze-cast hydroxyapatite scaffolds for bone tissue engineering applications. *Biomed Mater* 2008;3:025005.
45. Guda T, Walker JA, Singleton B, Hernandez J, Oh DS, Appleford MR, Ong JL, Wenke JC. Hydroxyapatite scaffold pore architecture effects in large bone defects *in vivo*. *J Biomater Appl* 2014;28:1016–1027.
46. Kim SS, Sun Park M, Jeon O, Yong Choi C, Kim BS. Poly(lactide-co-glycolide)/hydroxyapatite composite scaffolds for bone tissue engineering. *Biomaterials* 2006;27:1399–1409.
47. Kim SS, Ahn KM, Park MS, Lee JH, Choi CY, Kim BS. A poly(lactide-co-glycolide)/hydroxyapatite composite scaffold with enhanced osteoconductivity. *J Biomed Mater Res Part A* 2007;80:206–215.
48. Sun F, Zhou H, Lee J. Various preparation methods of highly porous hydroxyapatite/polymer nanoscale biocomposites for bone regeneration. *Acta Biomater* 2011;7:3813–3828.
49. Huang J, Xiong J, Liu J, Zhu W, Chen J, Duan L, Zhang J, Wang D. Evaluation of the novel three-dimensional porous poly (l-lactic acid)/nano-hydroxyapatite composite scaffold. *Biomed Mater Eng* 2015;26(Suppl. 1):S197–S205.
50. Leukers B, Gulkan H, Irsen SH, Milz S, Tille C, Schieker M, Seitz H. Hydroxyapatite scaffolds for bone tissue engineering made by 3D printing. *J Mater Sci Mater M* 2005;16:1121–1124.
51. Hollister SJ. Porous scaffold design for tissue engineering. *Nat Mater* 2005;4:518–524.

52. Elsinger EC, Leal L. Coralline hydroxyapatite bone graft substitutes. *J Foot Ankle Surg* 1996;35:396–399.
53. Xu Y, Wang DZ, Yang L, Tang HG. Hydrothermal conversion of coral into hydroxyapatite. *Mater Charact* 2001;47:83–87.
54. Roy DM, Linnehan SK. Hydroxyapatite formed from coral skeletal carbonate by hydrothermal exchange. *Nature* 1974;247:220–222.
55. Mygind T, Stiehler M, Baatrup A, Li H, Zou X, Flyvbjerg A, Kassem M, Bunger C. Mesenchymal stem cell ingrowth and differentiation on coralline hydroxyapatite scaffolds. *Biomaterials* 2007;28:1036–1047.
56. Oliveira JM, Silva SS, Malafaya PB, Rodrigues MT, Kotobuki N, Hirose M, Gomes ME, Mano JF, Ohgushi H, Reis RL. Macroporous hydroxyapatite scaffolds for bone tissue engineering applications: Physicochemical characterization and assessment of rat bone marrow stromal cell viability. *J Biomed Mater Res Part A* 2009;91:175–186.
57. Tamai N, Myoui A, Tomita T, Nakase T, Tanaka J, Ochi T, Yoshikawa H. Novel hydroxyapatite ceramics with an interconnective porous structure exhibit superior osteoconduction *in vivo*. *J Biomed Mater Res* 2002;59:110–117.
58. Huang J, Lin YW, Fu XW, Best SM, Brooks RA, Rushton N, Bonfield W. Development of nano-sized hydroxyapatite reinforced composites for tissue engineering scaffolds. *J Mater Sci Mater Med* 2007;18: 2151–2157.
59. Kim S, Kim SS, Lee SH, Eun Ahn S, Gwak SJ, Song JH, Kim BS, Chung HM. *In vivo* bone formation from human embryonic stem cell-derived osteogenic cells in poly(D,L-lactic-co-glycolic acid)/hydroxyapatite composite scaffolds. *Biomaterials* 2008;29:1043–1053.
60. Xiao X, Liu R, Huang Q. Preparation and characterization of nano-hydroxyapatite/polymer composite scaffolds. *J Mater Sci Mater Med* 2008;19:3429–3435.
61. Pitt CG, Chasalow F, Hibionada Y, Klimas D, Schindler A. Aliphatic polyesters. I. The degradation of poly (ε-caprolactone) *in vivo*. *J Appl Polym Sci* 1981;26:3779–3787.
62. Pitt G, Gratzl M, Kimmel G, Surles J, Sohindler A. Aliphatic polyesters. II. The degradation of poly(D,L-lactide), poly(ε-caprolactone), and their copolymers *in vivo*. *Biomaterials* 1981;2:215–220.
63. Zhao J, Guo LY, Yang XB, Weng J. Preparation of bioactive porous HA/PCL composite scaffolds. *Appl Surf Sci* 2008;255:2942–2946.
64. Heo SJ, Kim SE, Wei J, Hyun YT, Yun HS, Kim DH, Shin JW, Shin JW. Fabrication and characterization of novel nano- and micro-HA/PCL

composite scaffolds using a modified rapid prototyping process. *J Biomed Mater Res Part A* 2009;89:108–116.

65. Klar RM, Duarte R, Dix-Peek T, Dickens C, Ferretti C, Ripamonti U. Calcium ions and osteoclastogenesis initiate the induction of bone formation by coral-derived macroporous constructs. *J Cell Mol Med* 2013;17:1444–1457.
66. Evans SL, Gregson PJ. Composite technology in load-bearing orthopaedic implants. *Biomaterials* 1998;19:1329–1342.
67. Ducheyne P, Healy K, Hutmacher DE, Grainger DW, Kirkpatrick CJ. *Comprehensive Biomaterials: Online Version*. Newnes Elsevier: Elsevier; 2011.
68. Abu Bakar MS, Cheng MH, Tang SM, Yu SC, Liao K, Tan CT, Khor KA, Cheang P. Tensile properties, tension–tension fatigue and biological response of polyetheretherketone–hydroxyapatite composites for load-bearing orthopedic implants. *Biomaterials* 2003;24:2245–2250.
69. Wong KL, Wong CT, Liu WC, Pan HB, Fong MK, Lam WM, Cheung WL, Tang WM, Chiu KY, Luk KD, Lu WW. Mechanical properties and *in vitro* response of strontium-containing hydroxyapatite/polyetheretherketone composites. *Biomaterials* 2009;30:3810–3817.
70. Iskandar ME, Aslani A, Tian Q, Liu H. Nanostructured calcium phosphate coatings on magnesium alloys: Characterization and cytocompatibility with mesenchymal stem cells. *J Mater Sci Mater Med* 2015;26:189.
71. Furth ME, Atala A, Van Dyke ME. Smart biomaterials design for tissue engineering and regenerative medicine. *Biomaterials* 2007;28:5068–5073.
72. Bhardwaj G, Yazici H, Webster TJ. Reducing bacteria and macrophage density on nanophase hydroxyapatite coated onto titanium surfaces without releasing pharmaceutical agents. *Nanoscale* 2015;7:8416–8427.
73. Cheang P, Khor KA. Addressing processing problems associated with plasma spraying of hydroxyapatite coatings. *Biomaterials* 1996;17:537–544.
74. Wang BC, Lee TM, Chang E, Yang CY. The shear-strength and the failure mode of plasma-sprayed hydroxyapatite coating to bone— The effect of coating thickness. *J Biomed Mater Res* 1993;27:1315–1327.
75. Sun LM, Berndt CC, Gross KA, Kucuk A. Material fundamentals and clinical performance of plasma-sprayed hydroxyapatite coatings: A review. *J Biomed Mater Res* 2001;58:570–592.

76. Zheng XB, Huang MH, Ding CX. Bond strength of plasma-sprayed hydroxyapatite/Ti composite coatings. *Biomaterials* 2000;21:841–849.
77. Balani K, Anderson R, Laha T, Andara M, Tercero J, Crumpler E, Agarwal A. Plasma-sprayed carbon nanotube reinforced hydroxyapatite coatings and their interaction with human osteoblasts *in vitro*. *Biomaterials* 2007;28:618–624.
78. Xue W, Hosick HL, Bandyopadhyay A, Bose S, Ding C, Luk KDK, Cheung KMC, Lu WW. Preparation and cell–materials interactions of plasma sprayed strontium-containing hydroxyapatite coating. *Surf Coat Technol* 2007;201:4685–4693.
79. Lu YP, Li MS, Li ST, Wang ZG, Zhu RF. Plasma-sprayed hydroxyapatite+titania composite bond coat for hydroxyapatite coating on titanium substrate. *Biomaterials* 2004;25:4393–4403.
80. Ding SJ, Ju CP, Lin JHC. Morphology and immersion behavior of plasma-sprayed hydroxyapatite/bioactive glass coatings. *J Mater Sci Mater M* 2000;11:183–190.
81. Kar A, Raja KS, Misra M. Electrodeposition of hydroxyapatite onto nanotubular $TiO_2$ for implant applications. *Surf Coat Technol* 2006;201:3723–3731.
82. Jamesh M, Kumar S, Sankara Narayanan TSN. Electrodeposition of hydroxyapatite coating on magnesium for biomedical applications. *J Coat Technol Res* 2011;9:495–502.
83. Thanh DTM, Nam PT, Phuong NT, Que LX, Anh NV, Hoang T, Lam TD. Controlling the electrodeposition, morphology and structure of hydroxyapatite coating on 316L stainless steel. *Mater Sci Eng C* 2013;33:2037–2045.
84. Song YW, Shan DY, Han EH. Electrodeposition of hydroxyapatite coating on AZ91D magnesium alloy for biomaterial application. *Mater Lett* 2008;62:3276–3279.
85. Wang H, Eliaz N, Xiang Z, Hsu HP, Spector M, Hobbs LW. Early bone apposition *in vivo* on plasma-sprayed and electrochemically deposited hydroxyapatite coatings on titanium alloy. *Biomaterials* 2006;27:4192–4203.
86. Huang Y, Han SG, Pang XF, Ding QQ, Yan YJ. Electrodeposition of porous hydroxyapatite/calcium silicate composite coating on titanium for biomedical applications. *Appl Surf Sci* 2013;271:299–302.
87. Gopi D, Shinyjoy E, Sekar M, Surendiran M, Kavitha L, Kumar TSS. Development of carbon nanotubes reinforced hydroxyapatite

composite coatings on titanium by electrodeposition method. *Corros Sci* 2013;73:321–330.

88. Rojaee R, Fathi M, Raeissi K. Electrophoretic deposition of nanostructured hydroxyapatite coating on AZ91 magnesium alloy implants with different surface treatments. *Appl Surf Sci* 2013;285:664–673.
89. Kwok CT, Wong PK, Cheng FT, Man HC. Characterization and corrosion behavior of hydroxyapatite coatings on $Ti_6Al_4V$ fabricated by electrophoretic deposition. *Appl Surf Sci* 2009;255:6736–6744.
90. Besra L, Liu M. A review on fundamentals and applications of electrophoretic deposition (EPD). *Prog Mater Sci* 2007;52:1–61.
91. Zhitomirsky I, GalOr L. Electrophoretic deposition of hydroxyapatite. *J Mater Sci Mater M* 1997;8:213–219.
92. Javidi M, Javadpour S, Bahrololoom ME, Ma J. Electrophoretic deposition of natural hydroxyapatite on medical grade 316L stainless steel. *Mater Sci Eng C* 2008;28:1509–1515.
93. Aksakal B. The effect of immersion times on the solubility and bonding stress of hydroxyapatite coated $Ti_6Al_4V$ pedicle screws. *J Appl Biomater Biomech* 2010;8:33–41.
94. Wei M, Ruys AJ, Milthorpe BK, Sorrell CC, Evans JH. Electrophoretic deposition of hydroxyapatite coatings on metal substrates: A nanoparticulate dual-coating approach. *Journal of Sol–Gel Science and Technology* 2001;21:39–48.
95. Rath PC, Besra L, Singh BP, Bhattacharjee S. Titania/hydroxyapatite bi-layer coating on Ti metal by electrophoretic deposition: Characterization and corrosion studies. *Ceramics International* 2012;38: 3209–3216.
96. Bai Y, Neupane MP, Park IS, Lee MH, Bae TS, Watari F, Uo M. Electrophoretic deposition of carbon nanotubes — hydroxyapatite nanocomposites on titanium substrate. *Mater Sci Eng C* 2010;30: 1043–1049.
97. Alzubaydi TL, Alameer SS, Ismaeel T, Alhijazi AY, Geetha M. *In vivo* studies of the ceramic coated titanium alloy for enhanced osseointegration in dental applications. *J Mater Sci Mater Med* 2009;20 (Suppl. 1):S35–S42.
98. Mahmoodi S, Sorkhi L, Farrokhi-Rad M, Shahrabi T. Electrophoretic deposition of hydroxyapatite–chitosan nanocomposite coatings in different alcohols. *Surf Coat Technol* 2013;216:106–114.

99. Xiao X, Liu R, Tang X. Electrophoretic deposition of silicon-substituted hydroxyapatite/poly(epsilon-caprolactone) composite coatings. *J Mater Sci Mater Med* 2009;20:691–697.
100. Grandfield K, Sun F, FitzPatrick M, Cheong M, Zhitomirsky I. Electrophoretic deposition of polymer–carbon nanotube–hydroxyapatite composites. *Surf Coat Technol* 2009;203:1481–1487.
101. Boccaccini AR, Keim S, Ma R, Li Y, Zhitomirsky I. Electrophoretic deposition of biomaterials. *J R Soc Interf* 2010;7 (Suppl. 5): S581–S613.
102. Pang X, Zhitomirsky I. Electrophoretic deposition of composite hydroxyapatite–chitosan coatings. *Mater Charact* 2007;58:339–348.
103. Pang X, Casagrande T, Zhitomirsky I. Electrophoretic deposition of hydroxyapatite-$CaSiO_3$–chitosan composite coatings. *J Colloid Interf Sci* 2009;330:323–329.
104. Redepenning J, Venkataraman G, Chen J, Stafford N. Electrochemical preparation of chitosan/hydroxyapatite composite coatings on titanium substrates. *J Biomed Mater Res Part A* 2003;66A:411–416.
105. Tian Q, Liu H. Electrophoretic deposition and characterization of nanocomposites and nanoparticles on magnesium substrates. *Nanotechnology* 2015;26:175102.

# Chapter 4

# Tricalcium Phosphate-based Nanoceramics

**Kazuaki Hashimoto***

*Department of Life and Environmental Sciences, Chiba Institute of Technology, 2-17-1, Tsudanuma, Narashino, Chiba 275-0016, Japan*

**kazuaki.hashimoto@it-chiba.ac.jp.*

## Abstract

Tricalcium phosphate transforms into three polymorphs of $\beta$, $\alpha$, $\alpha'$, which change thermally from low to high temperatures. Alpha-tricalcium phosphate ($\alpha$-TCP) and beta-tricalcium phosphate ($\beta$-TCP) are widely known as bioceramics. $\alpha$-TCP is used in its powder form as a raw material for bone paste and bone cement. $\beta$-TCP is used as a granular, porous, and dense body in clinical applications, primarily as a bone-filling material for bone defect parts and as a substitute bone material for hard tissues. The characteristics of the crystal structures and principal manufacturing methods of tricalcium phosphates were described. Additionally, the solid solution substitution mechanisms of tricalcium phosphate ceramics for various metal ions and means for enhanced functionality of the obtained materials and applications to medical care were discussed as well. The improvement of functionality from solid solution substitution occurs primarily with the control of material properties such as thermal stability, solubility, and sinterability.

## 4.1 Introduction

The percentage of the population comprising individuals aged 65 years or older has been high among developed countries in recent years, as ageing of the population and declining birth rate are both progressing rapidly. There are also an increasing number of patients suffering from broken bones and osteoporosis due to declining bone density, which is associated with the advent of population ageing. Therefore, substitute bone materials for hard tissues used to fill and restore bone defect parts are becoming increasingly important. Among such materials, bioceramics that are made of calcium phosphate materials having the same properties as hard tissues have assumed an important position. Alpha-tricalcium phosphate ($\alpha$-$Ca_3(PO_4)_2$; $\alpha$-TCP) and beta-tricalcium phosphate ($\beta$-$Ca_3(PO_4)_2$; $\beta$-TCP) are widely known among such materials. $\beta$-TCP is used as a granular, porous, and dense body in clinical applications, primarily as a bone-filling material for bone defect parts and as a substitute bone material for hard tissues. $\alpha$-TCP is used in its powder form as a raw material for bone paste and bone cement. These varieties of calcium phosphates are highly bioactive and readily bond with bones in the living body. They are also absorbed in the living body and are gradually replaced by new bones, which is a characteristic that does not exist with other bioceramics.

## 4.2 History

Phosphoric acid is one of the three major elements used in fertilisers. It is not an overstatement to say that the development of calcium phosphate occurred hand in hand with chemical fertilisers since the birth of the calcium superphosphate industry in Germany, back in the 1840s.

Known phosphorus resources available in nature include phosphate minerals such as calcium phosphate minerals and phosphate band minerals.[1] Phosphate ores generally have a mineral composition of a solid solution made up of fluoroapatite and hydroxyapatite. Such phosphate minerals, however, include minerals of tricalcium phosphate (TCP) compositions such as the one referred to as whitlockite, which includes trace amounts of metal ions such as magnesium and iron.[2,3] This mineral was referred to as "whitlockite" in the past and this term was used as a mineral name for $\beta$-TCP, since the Debye–Scherrer pattern of this mineral was similar to that of $\beta$-TCP, which is a polymorphic substance of TCP. Studies in recent years revealed that whitlockite is a different substance from $\beta$-TCP in terms of chemical composition and thermal properties. $\beta$-TCP[4] and whitlockite have since then been treated as different substances, except for their crystal structures.[5]

On the other hand, a study on the pure crystal chemistry of TCP, which does not contain impurities, conducted in 1932 by Tromel revealed for the first time the existence of polymorphs of $\alpha$-$Ca_3(PO_4)_2$ ($\alpha$-TCP) and $\beta$-$Ca_3(PO_4)_2$ ($\beta$-TCP)[6] and the temperature for transition from $\beta$-TCP to $\alpha$-TCP was also reported to be 1125°C.[7] A new high-temperature phase was discovered in 1959 and it became evident that $\alpha'$-$Ca_3(PO_4)_2$ (super $\alpha$-TCP, $\alpha'$-TCP) could exist stably even at 1430°C or higher and could not be formed even when quenching to room temperature.[8] The formation of these phases will be discussed along with the CaO-$P_2O_5$ phase diagram. A high-pressure polymorphic substance ($\gamma$-TCP) is also known to exist.[9]

Hydroxyapatite (HAp; Ca/P = 1.67), which is an inorganic component of bones in living bodies, has been a subject of research as a bioceramic with biocompatibility since the 1970s. TCP is a calcium phosphate very similar to HAp and has been a subject of attention as a bioceramic with biodegradable properties since the 1990s. The solubility of TCP in a bioenvironment is greater than that of HAp. This means that while HAp is barely absorbed in living bodies, TCP is absorbed gradually and is transferred to the bones of the living body.

## 4.3 Process of $Ca_3(PO_4)_2$ Synthesis

### 4.3.1 Synthesis of TCP by solid-phase reaction method

Tricalcium phosphate ($Ca_3(PO_4)_2$; TCP) transforms into three polymorphs of $\beta$, $\alpha$, $\alpha'$, which change thermally from low to high

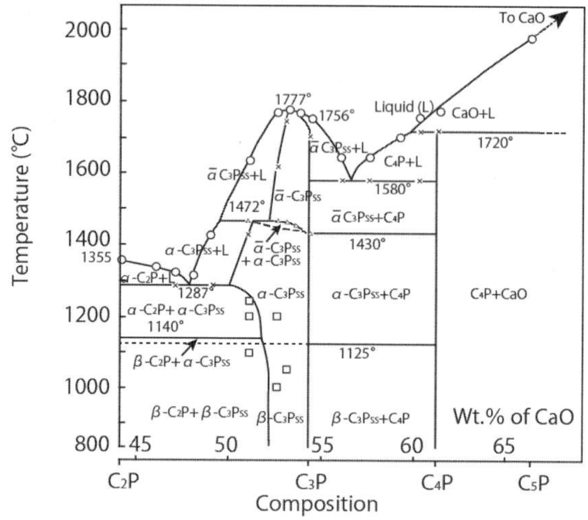


Figure 4.1 Phase diagram of $CaO–P_2O_5$ binary system.[7]
$C_2P$: $Ca_2P_2O_7$; $C_3P$: $Ca_3 (PO_4)_2$; $C_4P$: $Ca_4O (PO_4)_2$

temperatures. Figure 4.1 shows a binary system of the phase diagram for $CaO–P_2O_5$.

TCP is obtained by blending CaO and $P_2O_5$ in a molar ratio of 3:1 and then heating the mixture to trigger a mutual diffusional reaction in the solid state to cause a solid-phase reaction:

$$3CaO + P_2O_5 \rightarrow \beta\text{- or } \alpha\text{-}Ca_3(PO_4)_2 \tag{4.1}$$

In such cases, $\beta$-TCP is formed between 800°C and 1125°C, whereas $\alpha$-TCP is formed between 1125°C and 1430°C. This means that when manufacturing a raw material for bioceramics, $\beta$-TCP, which is stable at low temperatures, can be generally obtained by heating at 1000–1100°C, while $\alpha$-TCP, which is in the stabilised phase at high temperatures, can be obtained by first heating to the transition temperature or higher and then quenching to room temperature. This inhibits the $\alpha$–$\beta$ transition to obtain $\alpha$-TCP. Quenching of $\alpha'$-TCP, however, only achieves a metastable phase at room temperature, since the $\alpha'$–$\alpha$ transition is fast. Therefore, formation of $\alpha'$-TCP is difficult at room temperature. Furthermore, no formation of HAp from the solid-phase reaction under dry conditions was confirmed in this phase diagram.

### 4.3.2 Synthesis of TCP from ACP by precipitation method

Aqueous solutions of $Ca(NO_3)_2$ and $(NH_4)_2HPO_4$ (adjusted to pH 10) of prescribed concentrations are mixed while being agitated at 5°C or below, to precipitate amorphous calcium phosphate (ACP). The molar ratio of Ca/P as they are mixed is 1.50. After agitating for 5 min, the precipitate is washed with diluted ammonia water and then freeze-dried.

The compositions of the raw material solution and the product obtained by heating are shown in Figure 4.2.[10] An increased heating temperature changes the composition of the product from ACP to $\alpha$-TCP and then to $\beta$-TCP. Furthermore, the composition of the product changes from $\beta$-TCP to $\alpha$-TCP, due to the increased concentration of the raw material solution. $\alpha$-TCP is formed by heating at the transition temperature of 1125°C or higher, but this method

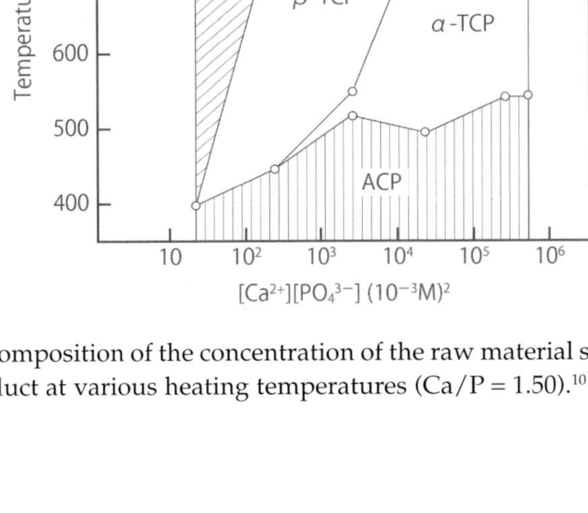


Figure 4.2 Composition of the concentration of the raw material solution and the obtained product at various heating temperatures (Ca/P = 1.50).[10]

makes it possible to synthesise $\alpha$-TCP at lower temperatures between 500°C and 800°C, with the prescribed concentration of the raw material solution.

### 4.3.3 Synthesis of TCP from gel using sol–gel method

A mixture of water, acetic acid, and ethanol is added to a mixture (Ca/P = 1.50) of calcium diethoxide in ethylene glycol and phosphite triester in ethanol to perform hydrolysis and polycondensation. The obtained sol is then gelated by drying at 80°C. The dried gel is heated at the $\alpha$–$\beta$ transition temperature or below to form $\beta$-TCP, while $\alpha$-TCP is formed when heating at the $\alpha$–$\beta$ transition temperature or higher.

### 4.3.4 Synthesis of TCP from DAp using mechanochemical method

$CaCO_3$ and $CaHPO_4 \cdot 2H_2O$ are blended in a molar ratio of 1:2 and then deionised water is added to the powder of this mixture to prepare a 10% slurry. This slurry is then placed inside a pot mill made of zirconium dioxide to mix and crush the powder while being rotated at 50 rpm. This is performed at room temperature or lower for 24 h, to cause a mechanochemical reaction. After the reaction occurs, the product is dried at 80°C. The composition of the product after the mechanochemical reaction is that of apatite with calcium defects (DAp; Ca/P = 1.50). DAp is heated at the $\alpha$–$\beta$ transition temperature or lower to form $\beta$-TCP, while $\alpha$-TCP is formed when heating at the $\alpha$–$\beta$ transition temperature or higher.

$$\begin{aligned} CaCO_3 + 2CaHPO_4 \cdot 2H_2O &\rightarrow \text{DAp (Ca/P} = 1.5) \\ &\rightarrow \beta\text{- or } \alpha\text{-}Ca_3(PO_4)_2 \end{aligned} \tag{4.2}$$

### 4.3.5 Preparation of $\beta$-TCP porous body

$\beta$-TCP of high purity is prepared, and then a surfactant and foam stabiliser are added to the prepared powder and dried. The powder is then baked to prepare a porous block body and porous granule body to manufacture artificial bones.

Next, an aqueous solution of polyacrylic acid ammonium, prepared as a 10% solution, is added to the $\beta$-TCP powder prepared to a granule diameter of 0.3 $\mu$m or smaller, and then an ultrasonic disperser is used to dissipate the powder in water. Polyoxyethylene nonyl phenyl ether is then added and foamed using an ultrasonic disperser. The creamy foam thus prepared is poured into a container for forming and dried at 40°C. The dried body is baked at 1000°C for 1 h.

The heated porous body is cut into a block body ($10 \times 10 \times 10$ mm) and granule bodies (2–5 mm in diameter) for use, as shown in Figure 4.3.

A scanning electron micrograph of the $\beta$-TCP porous body interior is shown in Figure 4.4. Furthermore, a micropore distribution curve of the $\beta$-TCP porous body is shown in Figure 4.5.

Characteristics of the manufactured porous body are the macropores measuring 50–400 $\mu$m and micropores measuring 0.8–2 $\mu$m that are accessible and such a porous property is believed to provide a good footing for cell spreading. The manufactured body features more macroscopic pores than autologous bones and

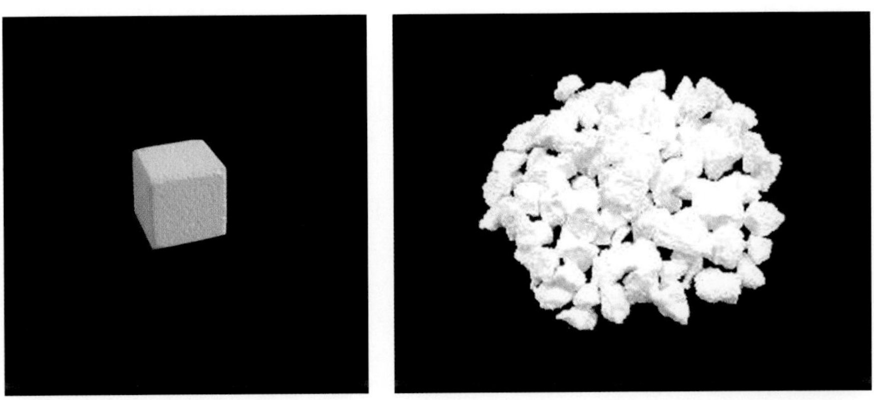

Figure 4.3 Photograph of external appearance of $\beta$-TCP porous body (left: block body; right: granule body).

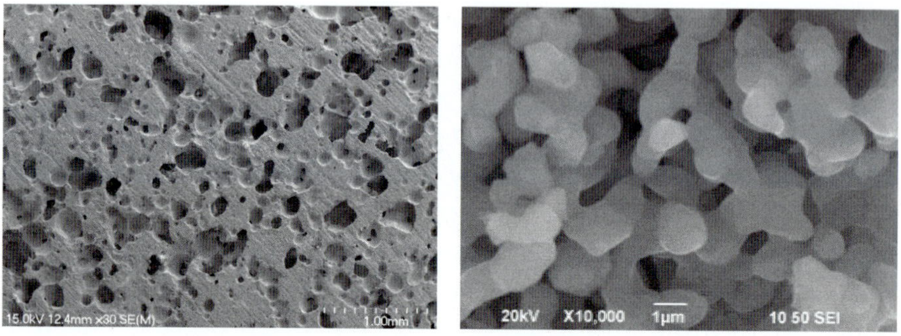


Figure 4.4 Scanning electron micrograph of the $\beta$-TCP porous body interior (left: 30 times magnification; right: 10,000 times magnification).

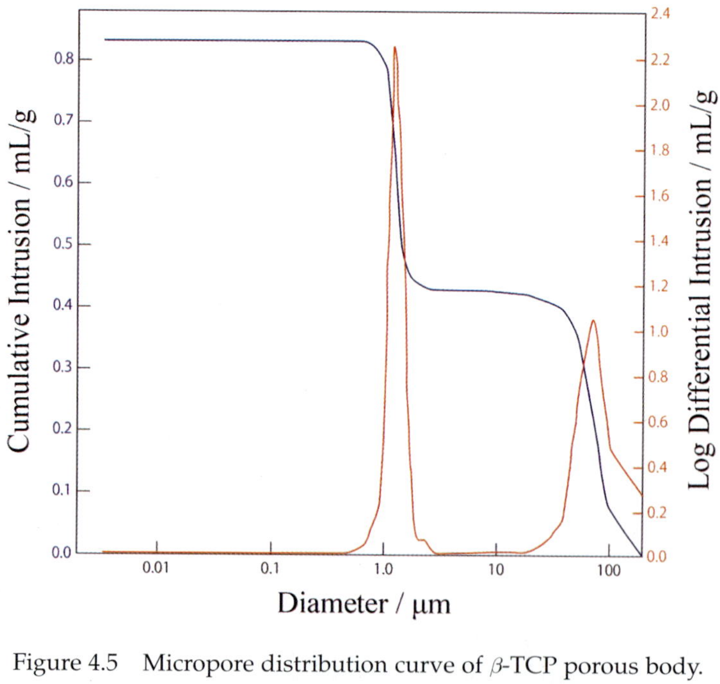


Figure 4.5 Micropore distribution curve of $\beta$-TCP porous body.

as such offers better blood flow and is adaptable to major bone ailments, which are ordinarily considered to have poor blood flow.

The porosity of the $\beta$-TCP porous body is 65–75% and the compressive strength is 2–3 MPa, which make the application of this body easy in surgical procedures. $\beta$-TCP also offers the ideal property of being a bone-filling material that is gradually replaced by the patient's own bone.

## 4.4 Enhanced Functionality of TCP by Solid Solution Substitution of Metal Ions

Hard tissues are comprised of calcium and phosphorus, but they also include electrolytic elements such as sodium, potassium, and magnesium, as well as trace amounts of base elements and quasi-base elements. Such base elements have been observed to grow in the body by bone remoulding and are expected to activate functions of various hard tissues and organs. Adding such base elements, therefore, is believed to enhance the biological functionality of bioceramics. Apart from studies on hard tissue constituents of calcium and phosphorus, many studies are being conducted on bioceramics with trace amounts of base elements or ions.

### 4.4.1 Composition control of $\alpha$-$Ca_3(PO_4)_2$ and mechanism of metal ion replacement

The crystalline structure of $\alpha$-TCP can be described as a monoclinic system with a space group of $P2_1/a$, with lattice constants of $a$ = 1.2887 nm, $b$ = 2.7280 nm, and $c$ = 1.5219 nm, a density of 2.86 g $cm^{-3}$ and $z$ = 24.[11] A projection diagram of the (0 0 1) surface of $\alpha$-$Ca_3(PO_4)_2$ is shown in Figure 4.6. A summary of the pseudo-apatite lattice is also shown in the figure, with a dotted line. An apatite structure and corresponding pseudo-apatite lattices are visible in the $\alpha$-TCP structure. The pseudo-apatite lattices are $b/3$ (0.91 nm) of the $b$-axis in the $\alpha$-TCP structure, which is equivalent to the $a$-axis length in the actual apatite structure that measures about 0.94 nm, while they are $c/2$ (0.76 nm) of the $c$-axis in the $\alpha$-TCP structure, which is equivalent to the $c$-axis length of the apatite structure that measures about 0.68 nm. Furthermore, the atomic arrangement of the $\alpha$-TCP crystal structure reveals that a column with an arrangement comprising of positive ions only

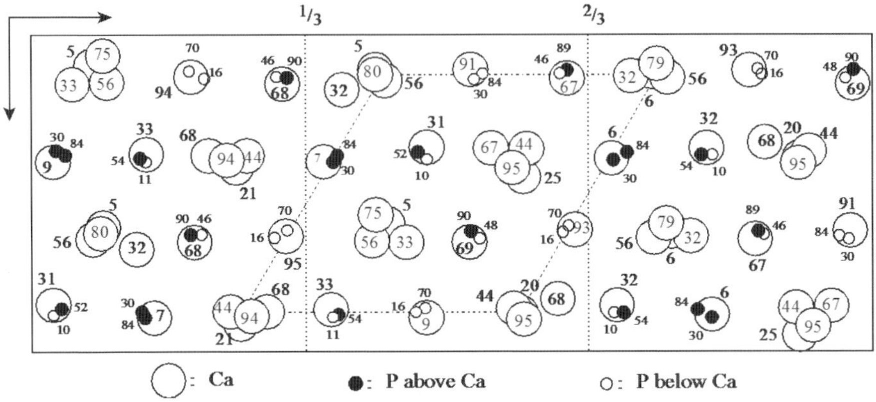


Figure 4.6 Projection diagram of the (0 0 1) surface of $\alpha$-$Ca_3(PO_4)_2$.[11]

(–Ca–Ca–Ca–Ca–) and that with an arrangement of positive and negative ions ($-PO_4-Ca-PO_4-\Box-PO_4-Ca-PO_4-\Box-PO_4-Ca-PO_4-$) reveals that a column with only positive ions exists in the same position as the Ca(1) position of the apatite structure. This column is the one present at the top of respective apatite lattices, but can turn into an apatite structure by replacing the positive ions with $OH^-$ ions or the like. On the other hand, the column with an arrangement comprised of positive ions and negative ions exists at the same position as the Ca(2) position of the apatite structure. This indicates that the atomic arrangement and crystalline structure of $\alpha$-TCP are similar to that of HAp.

The dimension of the unit lattice of $\alpha$-TCP is 0.180 $nm^3$ and is greater than that of $\beta$-TCP, which is 0.168 $nm^3$, suggesting that the former has a looser structure than the latter. Furthermore, since $\alpha$-TCP has a crystalline structure that is stable at higher temperatures, it is evident that the level of its internal energy is high. This confirms the quick compositional change that occurs from $\alpha$-TCP to apatite in water.

The study conducted by Sayer and his associates[12] is well known for the solid solution substitution of silicon in $\alpha$-TCP. Sayer revealed that stable TCP (Si-$\alpha$-TCP) is generated with silicon, when the molar ratio of $SiO_2$ and HAp is between 0.25 and 0.33 in the

HAp-$SiO_2$ system. The following reaction occurs when the $SiO_2$/HAp molar ratio is 0.25:

$$4Ca_5(PO_4)_3OH + SiO_2 \rightarrow 13/2Ca_{3+\delta}(P_{1-\delta}Si_\delta O_4)_2 + 2H_2O \qquad (4.3)$$

where $\delta$ is 1/13 in this instance. The chemical composition of Si-$\alpha$-TCP is $Ca_{3.08}(P_{0.92}Si_{0.08}O_4)_2$, which has an excessive number of calcium atoms. This composition can actually be expressed as $(12/13Ca_3(PO_4)_2 + 1/13Ca_4(SiO_4)_2$, however, with one of the 13 TCPs replaced by $Ca_4(SiO_4)_2$ to maintain overall charge compensation for the crystal.

Furthermore, the following reaction occurs when the $SiO_2$/HAp molar ratio is 0.33:

$$6Ca_5(PO_4)_3OH + 2SiO_2 \rightarrow 10Ca_3(P_{1-\gamma}Si_\gamma O_{4-\gamma/2})_2 + 3H_2O \qquad (4.4)$$

where $\gamma$ in this instance is 1/10 and the average chemical composition of Si-$\alpha$-TCP is $Ca_3(P_{0.9}Si_{0.1}O_{3.95})_2$, which has an oxygen atom missing when expressed in terms of the TCP composition. $Ca_3(P_{0.9}Si_{0.1}O_{3.95})_2$ can be described as a monoclinic system with a space group of $P2_1/a$, with lattice constants of $a$ = 1.28637 nm, $b$ = 0.91197 nm, and $c$ = 1.52327 nm, while $\beta$ = 126.37°. This crystalline structure roughly resembles the $\alpha$-TCP structure, with the $b$-axis length of the $\alpha$-TCP structure reduced to one-third.

Furthermore, the composition of the precipitate, derived with a constant molar ratio of 1.50 for Ca/(Si+P), with silicon added and the wet process used to synthesise the precipitate, was examined. Figure 4.7 shows the amount of additional silicon that would have an impact on the phase change of the product obtained by heating the derived precipitate at 1250°C. It is evident from the figure that a single phase of $\alpha$-TCP stabilised by silicon is formed with the added silicon amount of 0.59–1.14 wt.%, for such experimental conditions. The lattice constants for Si-$\alpha$-TCP derived by adding 0.87 wt.% Si are $a$ = 1.2874 nm, $b$ = 2.7372 nm, $c$ = 1.5225 nm, and $\gamma$ = 126.38°, which reveals that the structure is stabilised and resembles that of $\alpha$-TCP.

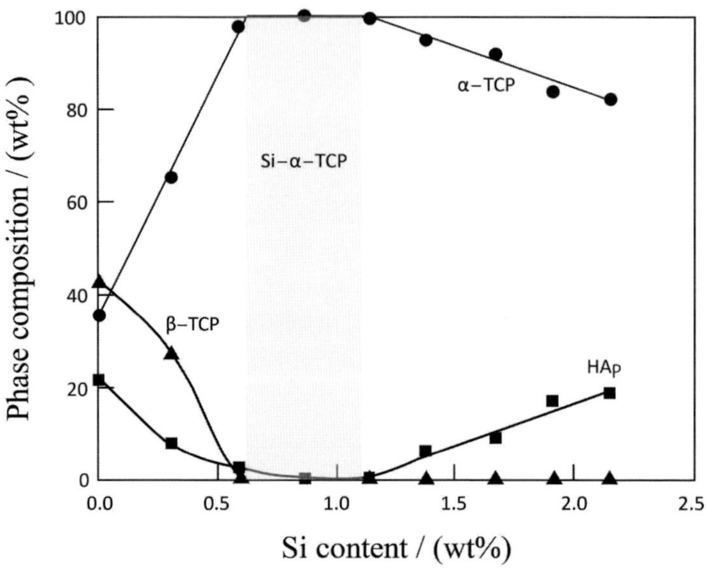


Figure 4.7 Amount of additional silicon that would have an impact on the phase change of the product obtained by heating the precipitate derived from the HAp-$SiO_2$ system at 1250°C.[12]

If bivalent positive ions such as magnesium and zinc ions are present in the calcium phosphate-type solid-phase reactions, Mg-$\beta$-TCP and Zn-$\beta$-TCP are readily formed, but in the case of silicon, Si-$\alpha$-TCP with the $\alpha$-TCP structure is formed in a stable manner.

### 4.4.2 Composition control of $\beta$-$Ca_3(PO_4)_2$ and mechanism of metal ion replacement

$\beta$-TCP belongs to the rhombohedral system, where the space group and the lattice filler are *R*3*c* (No. 161) and $Z = 21$ [calculated as $Ca_3(PO_4)_2$], respectively. Furthermore, the lattice constants for the hexagonal setting are $a = b = 1.04352$ nm, $c = 3.74029$ nm, $\alpha = \beta = 90°$, and $\gamma = 120°$.[13]

A schematic of the $\beta$-TCP crystalline structure is shown in Figure 4.8. Two columns, namely A and B columns, which are crystallographically independent, exist in the *c*-axis direction.

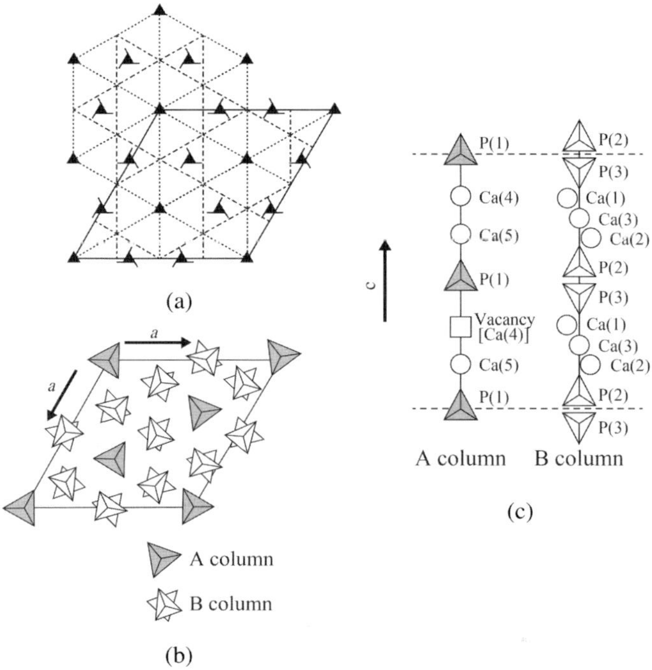


Figure 4.8 Schematic of the $\beta$-TCP unit cell. (a) Symmetry operation of (0 0 1)-plane of S.G. No. 161,[15] (b) structure as viewed down the [0 0 1]-axis, and (c) structure as viewed down the [1 0 0] axis.[13]

The A column exists on the threefold axis comprised of $–P(1)O_4–Ca(4)O_3–Ca(5)O_6–$, while the B column is comprised of $–P(3)O_4–Ca(1)O_7–Ca(3)O_8–Ca(2)O_8–P(2)O_4–$. Ca(1)–Ca(3) are arranged not linearly but in a deformed shape. The crystalline structure of $\beta$-TCP has an A column at the centre of the rotational axis. Six sets of B columns are arranged in a spiral rotation in the direction of [0 0 1], forming a tunnel structure. Ca(4) (site occupancy of 0.5) and Ca(5), which are crystallographically unique, exist in $\beta$-TCP and have an impact on the solid solution substitution of metal ions.

The change in the lattice constants of Mg-$\beta$-TCP, obtained by changing the blending quantity of magnesium ion, is shown in Figure 4.9.[14]

The X-ray diffraction pattern reveals that diffraction peaks that are matched with the diffraction geometry of $\beta$-TCP are observed

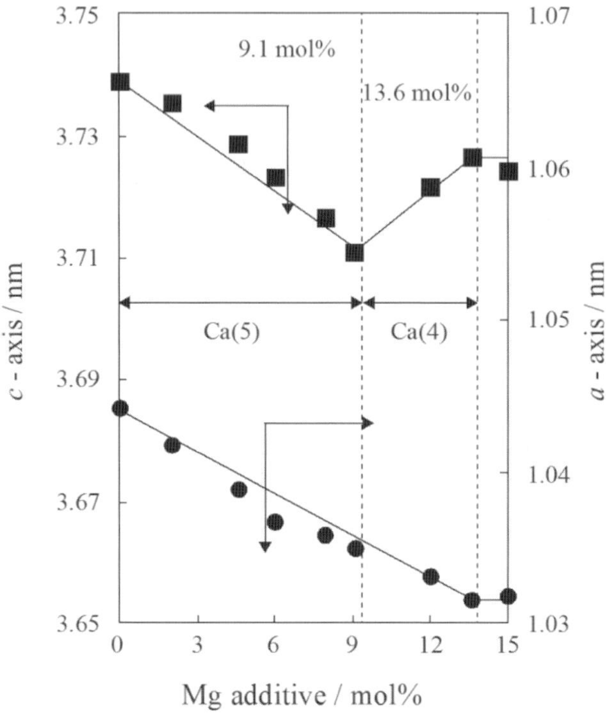


Figure 4.9 Change in lattice constants of Mg-β-TCP derived by changing blending quantity of magnesium ion.[14]

for each of the specimens. Furthermore, the diffraction peaks shifted towards the higher-angle side as the magnesium ion blending quantity increased, suggesting the phenomenon of solid solution substitution. The change in lattice constants in Figure 4.9 revealed that the *a*-axis length compressed linearly to 13.6 mol.%, while the *c*-axis length compressed linearly to 9.1 mol.% and then elongated to 13.6 mol.%. This indicates that the magnesium ion replaces Ca(5) of the A column with priority to 9.1 mol.% and then replaces Ca(4) only to 4.6 mol.% (site occupancy: 0.5) to replace a total amount of 13.6 mol.%. The chemical formulae of the products obtained from substitutions of 9.1 and 13.6 mol.% are $Ca_{19}Mg_2(PO_4)_{14}$ and $Ca_{18}Mg_3(PO_4)_{14}$, respectively. The solid solution substitution is similar to that of magnesium ion with zinc ion. In the case of β-TCP, the Ca site is prioritised where the solid solution substitution occurs, as described above.

Because research into solid solution substitution of metal ions into $\beta$-TCP has mainly centred on bivalent metal ions, there is not as much research on the formation of solid solutions by substitution with monovalent ions. Of the published work, the research by Ando *et al.*[16] into the two-component solid-phase reaction of $Ca_3(PO_4)_2$–$CaNaPO_4$ is well known. Additionally, the authors have reported the formation of solid solutions of $\beta$-TCP using the monovalent ions Li, Na, and K[14] and the thermal stability[17] and solubility[18] of these solid solutions.

When forming a solid solution substitution using monovalent metal ions, in addition to calcium phosphate, an alkali carbonate is used as the starting material and combined to produce a $(Ca+M^I+\square)/P$ molar ratio of 1.571. In this case, it is assumed that the monovalent positive ions will substitute with both the calcium ions and the vacant position. The molar ratio is expressed as $M^I = M^I/(Ca+M^I+\square)$. The synthesis method has the same solid-phase reaction as that previously described for the bivalent metal ions.

Figure 4.10 shows the change in the lattice constants for the samples obtained by changing the Na or K molar percent. As seen in the changes in the lattice constants, there were linear changes in each case with an accompanying increase in the molar percent of up to 9.1 mol.%. This observation corroborates the formation of partially substituted solid solutions. When substituting with Na ions, there was no change in the length along the *a*-axis of the solid solution; however, there was some contraction along the *c*-axis. On the other hand, for K ions, there was elongation of the *a*-axis of the solid solution and even greater contraction on the *c*-axis than with Na ions. Monovalent ion substitution forms solid solutions whereby the vacancy ($\square$) and Ca(4) of column A are substituted with the form $2M^I = Ca\ ion+\square$, at a solubility limit of 9.1 mol.%. This value of 9.1 mol.% is in agreement with the ratio of Ca(4) at each of the positions of the positive ions shown in Table 4.1. The chemical formula of the product at the solubility limit is $Ca_{20}M^I_2(PO_4)_{14}$ ($M^I$:Na,K).[19,20]

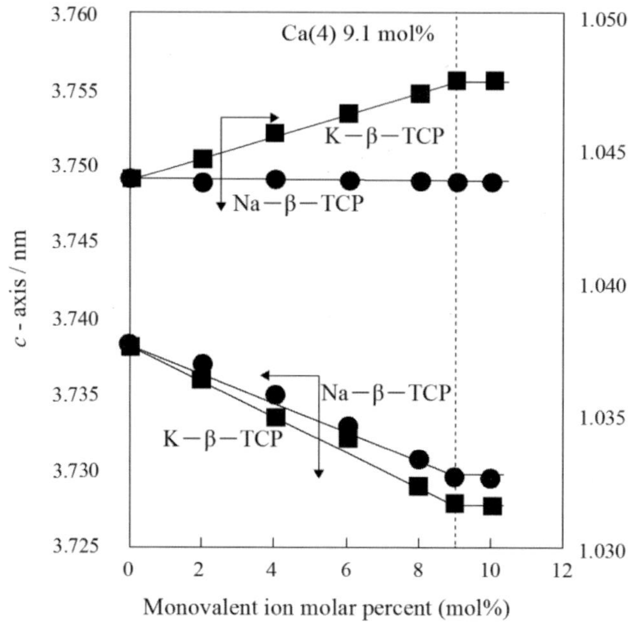


Figure 4.10 The lattice constants of Na-β-TCP and K-β-TCP obtained by different Na and K molar ratios.[14]

Table 4.1 Differences in crystal structures of β-TCP, Mg-β-TCP, and synthetic whitlockite.[25]

| | Atomic site | | | |
|---|---|---|---|---|
| | Cation site | | | |
| Chemical compound | Ca(1)–Ca(3) | Ca(4) | Ca(5) | Anion site |
| β-TCP, $Ca_3(PO_4)_2$ | $Ca_{18}$ | (Ca, □) | $Ca_2$ | $(PO_4)_{14}$ |
| $Mg_{9.1}$-β-TCP, $Ca_3(PO_4)_2$ with Mg 9.1 mol.% | $Ca_{18}$ | (Ca, □) | $Mg_2$ | $(PO_4)_{14}$ |
| $Mg_{13.6}$-β-TCP, $Ca_3(PO_4)_2$ with Mg 13.6 mol.% | $Ca_{18}$ | (Mg, □) | $Mg_2$ | $(PO_4)_{14}$ |
| Synthetic Mg-whitlockite | $Ca_{18}$ | $H_2$ | $Mg_2$ | $(PO_4)_{14}$ |

□: vacancy.

Furthermore, Wei *et al.*[21] showed that Si-$\beta$-TCP is derived by adding Zn and $SiO_2$ simultaneously, through silicon solid solution replacement of $\beta$-TCP. This mechanism causes $SiO_4$ solid solution substitution of the $PO_4$ site, which in turn causes the solid solution substitution of $Zn^{2+}$ ion at the Ca(5) site, which forms a six-coordinate structure with oxygen atoms. The theoretical value of BVS at the Ca(5) site is +2 in the case of $Ca^{2+}$, whereas the actual value of BVS at the Ca(5) site is close to about +3 due to the solid solution substitution of $Zn^{2+}$. This explains the electric charge compensation for the negative charge of the silicon solution, caused by excessive positive charge at this site.

### 4.4.3 Crystal structure of whitlockite

Whitlockite has a similar structure as $\beta$-TCP. It belongs to the rhombohedral system and the space group and the lattice filler are *R*3*c* (No. 161) and Z = 3 (calculated as $Ca_{18}M_2H_2(PO_4)_{14}$), respectively. The lattice constants of natural whitlockite ($Ca_{18.19}Mg_{1.17}Fe_{0.83}H_{1.62}(PO_4)_{14}$) discovered at the Palermo quarry are $a = b = 1.0330$ nm, $c = 3.7103$ nm, $\alpha = \beta = 90°$, $\gamma = 120°$ (hexagonal setting).[3] Furthermore, the lattice constants for the synthetic whitlockite ($Ca_{18}Mg_2H_2(PO_4)_{14}$) that includes $Mg^{2+}$ and $HPO_4^{2-}$ ions are $a = b = 1.0350$ nm, $c = 3.7085$ nm, $\alpha = \beta = 90°$, $\gamma = 120°$ (hexagonal setting).[5]

The crystalline structure of the synthetic whitlockite, similar to the structure of $\beta$-TCP, has an A column at the centre of the rotational axis. Six sets of B columns are arranged in spiral rotations in the direction of [0 0 1], forming a tunnel structure. The A column exists on the threefold axis comprised of $–P(1)O_4–H–MgO_6–$, while the B column is comprised of $–P(3)O_4–Ca(1)O_7–Ca(3)O_8–Ca(2)O_8–P(2)O_4–$. The structure of whitlockite has a hydrogen ion in Ca(4) and a magnesium ion in Ca(5) of the A column structure of $\beta$-TCP. It has the same structure when observed with an X-ray diffraction pattern as described above, but it is evident that it is crystallographically different (Table 4.1).

Whitlockite had been studied primarily in the field of mineralogy, but there have also been research reports in the medical and dental fields. Vahl *et al.*[22] as well as Johnson *et al.*[23] reported in their studies on biological hard tissues and dental cavities with abnormal calcifications in humans and animals that whitlockite had been formed in hard tissues, based on their observations using X-ray diffraction and electron microscopy. Furthermore, Rowles[24] investigated the amount of magnesium contained in dental calculus and the amount of whitlockite formed and they observed that when there is at least 2 wt.% of magnesium in the dental calculus, whitlockite is entirely formed. No studies using synthetic whitlockite as a biomaterial have been reported so far.

### 4.4.4 Thermodynamic behaviour of metal ion-substituted TCP solid solutions

We also focused on the thermodynamic behaviour of the ions substituted into the $\beta$-TCP structure. Because the substituted ions are sufficiently isolated from the other positive ions in the crystal, as long as the charge of the substituted ion is the same, the non-electrostatic repulsion of the M–O and M–M bonds will be roughly the same across the entire crystal. However, because the electrostatic energy of the substituted M–O bond is proportional to $1/r$ (where $r$ is the radius of the ion), if the M–O bond length contracts, as in Mg ion substitution, the electrostatic bond strengthens and the internal energy of the crystal ($U$) is reduced. Additionally, substituting with Mg ions increases the bond entropy ($S$) at the Ca positions. In other words, $U$ (Mg-$\beta$-TCP) < $U$ ($\beta$-TCP), and $S$ (Mg-$\beta$-TCP) > $S$ ($\beta$-TCP). If this is considered with the introduction of the Helmholtz energy ($F$) equation, $F = U - TS$, then $F$ ($\beta$-TCP) > $F$ (Mg-$\beta$-TCP), confirming that Mg-$\beta$-TCP is a thermodynamically stable product.

### 4.4.5 Solubility of metal ion-substituted TCP solid solutions

The solubility of inorganic compounds (solubility product: $K_{sp}$) depends on the relationship $\Delta G° = RT \ln K_{sp}$. This means that when the high-temperature stability of the $\beta$-TCP structure is high, the solubility declines, whereas when the high-temperature stability is low, the solubility improves. In other words, it is possible to estimate the solubility of β-TCP with substituted metal ions, based on the thermal stability evaluation of the β–α transition. In the case of α-TCP, the crystallographic stability decreases with $Mg^{2+}$, $Zn^{2+}$, and $SiO_4^{4-}$ ion solutions, while the solubility increases. On the other hand, in the case of β-TCP, the crystallographic stability increases with $Mg^{2+}$ and $Zn^{2+}$ solutions and the solubility decreases. It is therefore possible to control both the thermal stability and solubility in this manner.

Figure 4.11 shows the solubility of β-TCP and metal ion-substituted β-TCP solid solutions. Thus, a fixed amount of a powdered sample was suspended in 0.08 M acetic acid–sodium acetate buffer solution (0.10 g powder per 30 $cm^3$ solution) with a pH of 5.5, and its solubility was evaluated using a calcium electrode.[18] The 9.1 mol.% Sr, Mg, and Na ion-substituted β-TCP solid solutions

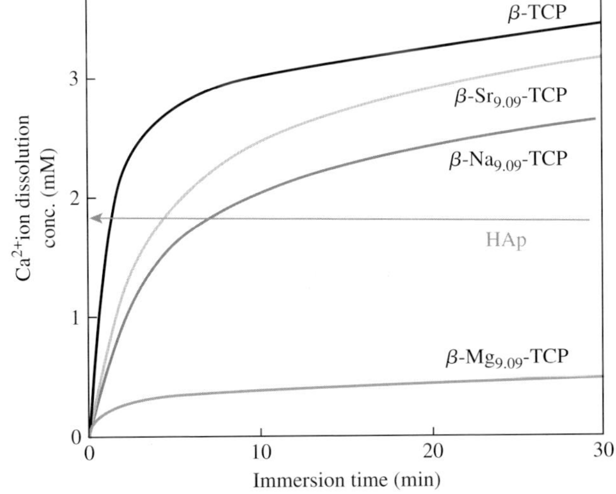


Figure 4.11 Evaluation of the solubility of solid solutions of each metal-ion-substituted β-TCP in acetic acid–sodium acetate buffer solution.[18]

were used as samples. The Ca(1)–Ca(3) site of the $\beta$-TCP is substituted by Sr ion, the Ca(5) site of the $\beta$-TCP is substituted by Mg ion, and the Ca(4) site of the $\beta$-TCP is substituted by Na ion. The results showed that the solubility for each metal-ion-substituted solid solution decreased in the order of $\beta$-TCP > bivalent [Ca(1)–(3) substitution] > monovalent [Ca(4) substitution] > HAp > bivalent [Ca(5) substitution]. Additionally, the solubility decreased proportionally with increasing amounts of substitution. The solubility of solid solutions of $\beta$-TCP corresponds with the thermal stability and, crystallographically, the structural stability of the 6-coordinated Ca(5), contributes to this thermal stability.

## 4.5 Application to Medical Care

Porous $\beta$-TCP, which is one of the artificial bone-filling materials, has osteoconductive characteristics, biocompatibility, and biodegradable characteristics. It has been used in many clinical studies and its usefulness has been confirmed. However, the effective materials for bone hyperplasia are not yet established. A transplant of the autologous bone is the first choice in osteoplasty of the laterality outside a bone. In addition to the problem of insult from the donor side, satisfactory bone hyperplasia is difficult to accomplish because there are many variables in the shape of the graft, quantity of absorption, etc. The conventional bone-filling material is mixed with the autologous bone that plays a supporting role. From the result shown in Figure 4.12, it was understood that the $\beta$-TCP porous block body as a biodegradable material is replaced by autologous bone 6 months after surgery.[26] The $\beta$-TCP porous block body showed high remodelling activity while maintaining the shape of the block and showed substitution of the bone sequentially over time. It was suggested that it was effective for a clinically difficult bone hyperplasia of the laterality.

The improvement of functionality from solid solution substitution occurs primarily with the control of material properties such as solubility and sinterability of the solid solution. If the concentration of calcium ions dissolved from a given material can be

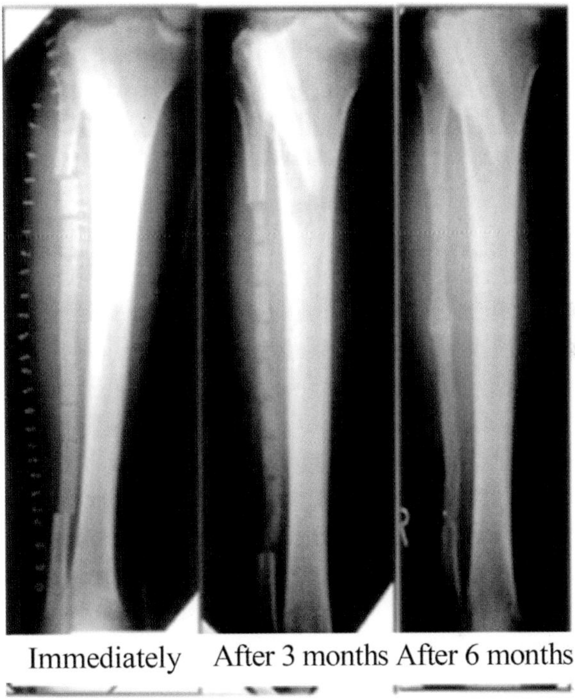


Figure 4.12 Filling of right fibular defect after harvest for autograft in scoliosis surgery.[26]

inhibited, cell adhesion by integrin is believed to accelerate, which in turn makes the early replacement by autologous bone possible.

Furthermore, there are essential elements in a living body that maintain homeostasis and facilitate normal metabolism. Metabolism of cells in skeletal formation is important for hard tissues such as bones. Electrolytic ions such as calcium, magnesium, sodium, and potassium are activated by cell metabolism. In bone cells, sodium ions activate bone absorption of osteoclasts and therefore act on the bone metabolism that is initiated by the activation of osteoclasts. Magnesium ions act on the calcification process as well as vulnerability of bones and mineral metabolism. Strontium ions influence the growth of bones and reduction of bone absorption. Iron ions are one of the trace ions that constitute hard tissues and play an important role in the function of many proteins in the circulatory system. Zinc ions are trace essential ions that provide an inhibiting effect on bone absorption by osteoclasts and similarly promote formation of bones. In particular, they are responsible for activating aminoacyl-tRNA synthetase, which increases the synthesis of osteocalcin in osteoblasts, as well as proteins that include production of IGF-I and TGF-$\beta$, to activate the osteoblasts. Silicon ions act on the metabolic mechanism that contributes to osteogenesis and the metabolic process that is involved with the manifestation of bone cells and cell junctions. Each ion described above is expected to provide pharmacological effects on bone metabolism, but there have not been many studies to evaluate them experimentally.

Ito *et al.*[27] mention that other than the usual melting of materials, Zn-$\beta$-TCP that includes a minute amount (about 1 wt.%) of zinc provides a rapid reduction in reabsorption due to the sedative action of Zn-$\beta$-TCP with respect to the formation of osteoclasts or activation of osteoclasts. They performed a cell evaluation using the mesenchymal cells of rats (MC3T3-E1) and demonstrated that Zn-$\beta$-TCP increases cell growth and activates ALP. Figure 4.13 shows the influence of Zn content in ceramics on the relative proliferation ratio. It was accepted that proliferation of cells on $\beta$-TCP

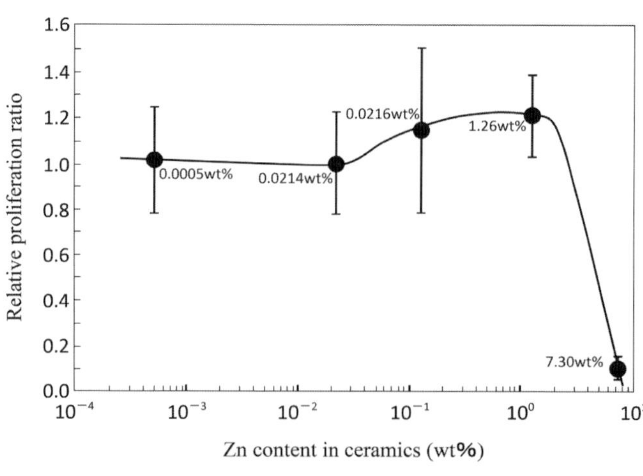


Figure 4.13 Influence of Zn content in ceramics to relative proliferation ratio.[27]

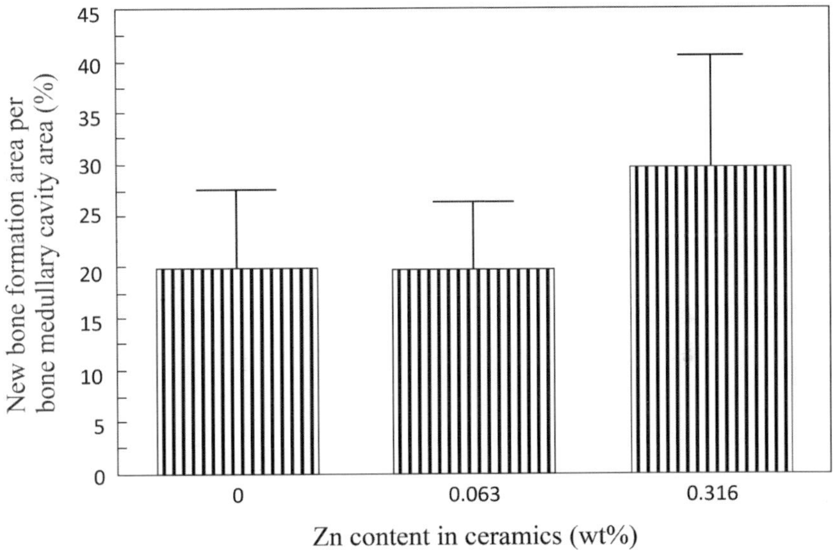


Figure 4.14 Influence of Zn content in ceramics to new bone formation area per bone medullary cavity area on the rabbit thighbone.[28]

ceramics substituted by 0.0216–1.26 wt.% zinc content occurred. However, the cell proliferation decreased sharply on the $\beta$-TCP ceramics substituted by 7.30 wt.% zinc content. From this, it is understood that there is an optimum amount of the additive ion.

Furthermore, an *in vivo* evaluation using rabbits found that Zn-$\beta$-TCP increased the amount of new bones formed by about 50% in 4 weeks, as shown in Figure 4.14 and proved that the zinc ion in Zn-$\beta$-TCP has a pharmacological effect for promoting bone formation.[28]

## 4.6 Summary

Alpha-tricalcium phosphate [$\alpha$-$Ca_3(PO_4)_2$; $\alpha$-TCP] and beta-tricalcium phosphate [$\beta$-$Ca_3(PO_4)_2$; $\beta$-TCP] are widely known to have osteoconductive, biodegradable, and biocompatibility characteristics. In this chapter, the history, principal manufacturing methods, and characteristics of the crystal structures (including their obtained forms) of tricalcium phosphates were described. Additionally, the solid solution substitution mechanisms of tricalcium phosphate ceramics for various metal ions, as well as means for enhanced functionality of the obtained materials and applications to medical care were discussed as well. The improvement of functionality from solid solution substitution occurs primarily with the control of material properties such as thermal stability, solubility, and sinterability.

Descriptions of the crystal structures of various calcium phosphates that are used in clinical applications as biomaterials, as well as the substitution mechanism of trace metals, which are inorganic biological constituents and trace constituents of bones in various calcium phosphates, along with the changes in their physical properties, were provided in this chapter. Biomaterials subject to composition control by solid solution substitution are affected by the solubility and sinterability of biomaterials derived from atomic positions that are replaced in such structures, as well as the amount of substitutions. These are furthermore considered to have an impact on bone absorption and bone neoplasia and even on the dynamic characteristics of biomaterials. Moreover, pharmacological effects of the substitution of various metal ions on calcium phosphate-based biomaterials are expected to become research topics for the future. Along with such composition control, combinations of physiologically active substances such as inductive differentiating factors and cells are considered promising biomaterials of the next generation.

## Acknowledgements

This study was supported in part by a grant-in-aid (No. 25350554) from the Ministry of Education, Science, Sports, and Culture of Japan.

## References

1. Kanazawa T. (ed.). *Inorganic Phosphate Materials*, Materials Science Monograph 52. Tokyo: Koudansha-Elsevier; 1989; pp. 8–13.

2. Frondel C. Whitlockite: A new calcium phosphate, $Ca_3(PO_4)_2$. *Am Miner* 1941;26:145–152.
3. Calvo C, Gopal R. The crystal structure of whitlockite from the Palermo Quarry. *Am Miner* 1975;60:120–133.
4. Schroeder LW, Dickens B, Brown WE. Crystallographic studies of the role of Mg as a stabilizing impurity in $\beta$-$Ca_3(PO_4)_2$. II. Refinement of Mg-containing$\beta$-$Ca_3(PO_4)_2$. *J Solid State Chem* 1977;22:253–262.
5. Gopal R, Calvo C, Ito J, Sabine WK. Crystal structure of synthetic Mg-whitlockite, $Ca_{18}Mg_2H_2(PO_4)_{14}$. *Can J Chem* 1974;52:1155–1164.
6. Tromel G. *Beitrage zur Kenntnis des Systems Kalziumoxyd-Phosphorpentoxd*, Vol. 14, Dusseldorf: Mitt Kaiser-Wilhelm-Inst Eisenforschg; 1932; pp. 25–34.
7. Welch JH, Gutt W. High-temperature studies of the system calcium oxide-phosphorus pentoxide. *J Chem Soc* 1961;1961:4442–4444.
8. Nurse RW, Welch JH, Gutt W. High-temperature phase equilibria in the system dicalcium silicate–tricalcium phosphate. *J Chem Soc* 1959; 1959:1077–1083.
9. Roux P, Louer D, Bonel G. Sur une nouvelle forme Cristalline du phosphate tricalcique. *Compt Rend Acad Sci Ser C* 1978;286:549–551.
10. Umegaki T, Shiba T, Kanazawa T. Thermal change of amorphous calcium phosphate. Effect of preparative conditions. *J Ceram Soc Jpn* 1984;92(11):612–620.
11. Mathew M, Schroeder LM, Dickens B, Brown WE. The crystal structure of $\alpha$-$Ca_3(PO_4)_2$. *Acta Cryst B* 1977;33:1325–1333.
12. Sayer M, Stratilatov AD, Reid J, Calderin L, Stott MI, Yin X, MacKenzie M, Smith TJN, Hendry JA, Langstaff SD. Structure and composition of silicon-stabilized tricalcium phosphate. *Biomaterials* 2003;24:369–382.
13. Yashima M, Sakai A, Kamiyama T, Hoshikawa A. Crystal structure analysis of $\beta$-tricalcium phosphate $Ca_3(PO_4)_2$ by neutron powder diffraction. *J Solid State Chem* 2003;175:272–277.
14. Yoshida K, Hyuga H, Kondo N, Kita H, Sasaki M, Mitamura M, Hashimoto K, Toda Y. Substitution model of monovalent (Li, Na, and K), divalent (Mg), and trivalent (Al) metal Ions for $\beta$-tricalcium phosphate. *J Am Ceram Soc* 2006;89:688–690.
15. Hahn Th., *International Tables for Crystallography*, Volume A, Fourth Edition, Dordrecht/Boston/London: Kluwer Academic Publishers; 1995; pp. 522–525.

16. Ando J, Matsuno S. $Ca_3(PO_4)_2$–$CaNaPO_4$ system. *Bull Chem Soc Jpn* 1968;41:342–347.
17. Matsumoto N, Yoshida K, Hashimoto K, Toda Y. Thermal stability of β-tricalcium phosphate doped with monovalent metal ions. *Mater Res Bull* 2009;23:1889–1894.
18. Matsumoto N, Yoshida K, Hashimoto K, Toda Y. Dissolution mechanisms of β-tricalcium phosphate doped with monovalent metal ions. *J Ceram Soc Jpn* 2010;118:451–457.
19. Quillard S, Paris M, Deniard P, Gildenhaar R, Berger G, Obadia L, Bouler J-M. Structural and spectroscopic characterization of a series of potassium- and/or sodium-substituted β-tricalcium phosphate. *Acta Biomater* 2011;7:1844–1852.
20. Morozov VA, Belik AA, Kotov RN, Presnyakov IA, Khasanov SS, Lazoryak BI. Crystal structures of double calcium and alkali metal phosphates $Ca_{10}M(PO_4)_7$ (M = Li, Na, K). *Crystallogr Rep* 2000;45:13–20.
21. Wei X, Ugurlu O, Ankit A, Acar HY, Akinc M. Dissolution behavior of Si,Zn-codoped tricalcium phosphates. *Mater Sci Eng C* 2009;29: 126–135.
22. von Vahl J, Hohling HJ, Frank RM. Elektronenstrahlbeugung an rhomboedrisch aussehenden Mineralbildungen in kariösem Dentin. *Arch Oral Biol* 1964;9:315–320.
23. Johnson NW, Taylor BR, Berman DS. The response of deciduous dentine to caries studied by correlated light and electron microscopy. *Caries Res* 1969;3:348–368.
24. Rowles SL. The precipitation of whitlockite from aqueous solutions. *Bull Soc Chem Fr* 1968:1797–1802.
25. Elliott JC. *Structure and Chemistry of the Apatites and Other Calcium Orthophosphates*. Amsterdam: Elsevier; 1994; pp. 37–42.
26. Nakajima H, Arai E, Sugiura H, Katagiri H, Nishida Y. Osteoanagenesis of the defective part of bone tumor and fibula collection part. *Med Asahi* 2003;32(9):8–11 (in Japanese).
27. Ito A, Kawamura H, Otsuka M, Ikeuchi M, Ohgushi H, Ishikawa K, Onuma K, Kanzaki N, Sogo Y, Ichinose N. Zinc-releasing calcium phosphate for stimulating bone formation. *Mater Sci Eng C* 2002;22: 21–25.
28. Kawamura H, Ito A, Miyakawa S, Layrolle P, Ojima K, Ichinose N, Tateishi T. Stimulatory effect of zinc-releasing calcium phosphate implant on bone formation in rabbit femora. *J Biomed Mater Res* 2000;50(2):184–190.

# Chapter 5

# Calcium Phosphate Nanocoatings: Production, Physical and Biological Properties, and Biomedical Applications

**Besim Ben-Nissan* and Andy H. Choi†**

*School of Mathematical and Physical Sciences, Faculty of Science, University of Technology Sydney, Australia*

*besim.ben-nissan@uts.edu.au

†ahchoi@hotmail.com

**Abstract**

Over the years, the research and applications of calcium phosphate materials as nanocoatings for orthopaedic and dental implants and medical devices have undergone a revolution from being a curiosity to being a state-of-the-art technique to improve osteointegration. Nanocoatings, nanolaminates, and their composites offer the possibility of modifying the surface properties of surgical-grade materials to achieve improvements in *in vivo* reliability and performance as well as increased protection from the release of harmful or unnecessary metal ions through proper selection of coating materials. They can also be produced to deliver exceptional bioactivity and faster tissue-bonding properties due to their increased surface area and their nanocrystalline structure which are inter-related. This chapter aims to provide a brief review on the recent advances in calcium phosphate nanocoatings specifically on sol–gel-derived coating and their related nanocomposites as surface modification methods for dental and orthopaedic applications.

## 5.1 Introduction

Nanocoatings are thin films with thicknesses below the range of micron-sized coatings (<1 $\mu$m) that may contain homogeneous and isotropic compounds. Usual thickness is below 100 nm for a single coating. Multiple layers can be easily produced with appropriate chemical, physical, mechanical, and biological properties. They can be applied with different compositions as nanolaminates or multi-layered gradient coatings. Advantages of nanocoatings include the ease of synthesis, low cost due to the small amounts of materials needed for thin coating processes, range of application methods, purity due to the choice of raw materials, and ability to be mixed with other nanoparticles or compounds.[1–4]

Orthopaedic and maxillofacial implants that are currently being employed are manufactured with metallic materials such as bioinert cobalt chromium alloys or titanium alloys and do not chemically bond to bone unless modified. For this reason, they are surgically inserted and bonded with bone cements or by mechanical fixation. An alternative approach is to use micro-textured or surface-modified structures which allow mechanical interlocking and in certain instances chemical bonding with the bone. This can be achieved by fixing porous surfaces manufactured in such a manner that it possesses suitable surface pores or by micro- or macro-texturing using beads or wires. The surface area increases as a result of these structures and hence the fixation.[1]

After an implantation, a number of tissue responses can occur at interfaces between the implanted biomaterial and hard or soft tissues. Past studies and a number of excellent review papers[5–8] proposed five possible *in vivo* interfacial tissue responses at the tissue/implant interface: (a) if the material is non-toxic (biocompatible) and biodegradable, the surrounding tissue will replace it; (b) if the material is biocompatible but biologically inactive, a fibrous tissue might form around the implant that eventually may facilitate inflammation and loosening; (c) if the material is non-toxic and biologically active, an interfacial bond forms between the material and the tissue interface (osteointegration); (d) if the material is non-toxic but bioinert, a weak bond might form (ankyloses) or it can be surgically bonded using synthetic external cements (bone cements); and finally (e) if the implant material is toxic, the surrounding tissue dies.

The preventions of inflammatory responses to orthopaedic and maxillofacial implants are vital prerequisites regardless of the design or the implant materials used. Furthermore, an implant is also required to promote osteointegration and sound adequate biological fixation. The purpose of applying either micro- and/or nanocoatings to these implant materials is to improve bioactivity and, if possible, protect that material against degradation and the release of metal ions. Eventually, this leads to an improved structure and environment for the growth of new bone.[1–3]

A number of factors can influence the success rate of an orthopaedic or maxillofacial implant: the materials used and their structure and properties; the design of the implant or device; biomechanics and application and influence of the functional loading; patient's health and medical condition and last but not the least, the surgical procedure or technique employed.

During the last three decades, concerns and issues related to adhesion and biological interactions to increase longevity and reliability of implant systems have motivated the exploration of surface modification towards early osteointegration, bone-implant adaptability, and rapid healing. Improvement of surfaces by increasing bioactivity with chemical and biological means and surface macro- and micro-texturing has been the main focus for a large number of researchers in the biomedical and surgical field.

On the other hand, the microstructure and properties of nanostructured materials such as nanocoatings are governed by factors such as their processing or synthesis method, chemical composition, thickness of the coatings, and their structure. Techniques such as physical and chemical vapour depositions, electrochemical vapour deposition, metal-organic chemical vapour deposition, thermal or diffusion conversion, plasma and fusion coating, and sol–gel have been used to deposit coatings on both in the micro- and nanoscale.

## 5.2 Calcium Phosphate Nanocoatings

Hydroxyapatite (HAp) is widely accepted as a biocompatible material that exhibits the chemical resemblance of the mineral component of bone and teeth.[6,9] Most published information on HAp is classified under calcium phosphate, to which HAp belongs. Calcium phosphates are classified by particular solubilities within the physiological environment and their bonding mechanisms to the surrounding tissues, and hence by their ability to degrade.[10,11] As calcium phosphate (or HAp) comes into contact with body fluid, its surface ions can be exchanged with those of the aqueous solution. Alternatively, various ions and molecules such as collagen and proteins can be adsorbed onto their surface to produce biological coatings.[2,3]

Although bulk calcium phosphates have been used for surgical applications since the 1920s,[12] calcium phosphate macro- and micro-coatings are relatively recent and were utilised as porous coating materials since the early 1980s.[13] The primary purpose of using calcium phosphate as a bioactive coating is to establish a rapid and strong biological attachment to the soft or hard tissue. The definition of biological fixation can be described as the process in which the implant or prosthetic component bonds firmly to the host bone through bone in-growth with or without the use of mechanical fixation but without using any adhesive. The advantage of using HAp coatings lies in the supply of calcium and phosphate ions which may stimulate and/or facilitate the growth of new bone tissue on and toward the implant. However, a good balance must be reached between the solubility of the coating and the rate of bone growth to achieve mechanical integrity under functional loading. This would permit adequate mechanical bonding and properties at the implant–tissue interface for long-term survival of an implant.[3,4]

For successful clinical outcome of a biocompatible coated implant, it has been suggested that the ideal environment for bone growth is achieved when the metal ion release is reduced, good mechanical interlocking is created, and a bioactive surface is available for chemical and biological bond.[5] Quality calcium phosphate-coated implants induce rapid healing and attach more completely to the bone tissue.[1–3]

Properties of coatings and specifically nanocoatings depend on their thickness, constituent phases, and their crystallinity which determine their long-term performance. Animal models and human trials have shown that thin-film depositions of HAp and other calcium phosphates on implant surfaces accelerated early bone formation and increased the bond strength between the implant and bone.[13–15] In an oral implant study, it was shown that an increase in osteoconduction along the calcium phosphate treat surfaces was observed during the first 2 months of post-implantation.[13] When they examined the effects of discrete crystalline deposition of calcium phosphate nanoparticles added to dual acid-etched dental implant surfaces placed in the human posterior maxilla, they observed excellent osteointegration. They suggested that the nanometric particle deposition of calcium phosphate crystals can be advantageous for shortening the implant healing period, providing earlier fixation.[13]

In 2011, a study by Jimbo *et al.*[14] has provided further evidence which supported the fact that nanocalcium phosphate-coated surface enhances the osteointegration process. Using a rabbit model, they compared the conventional removal torque evaluation and gene expression in tissues around nanostructured calcium phosphate-coated implants via real-time reverse transcription with those of uncoated implants. Twenty threaded titanium alloy dental implants, half prepared with stable HAp nanoparticle surface and half grit-blasted, acid-etched, and heat-treated, were inserted into rabbit femurs. Preoperatively, the implants were morphologically and topographically characterised. After 3 weeks of healing, the samples were retrieved for histomorphometry. The nanomechanical properties of the surrounding bone were evaluated using nanoindentation. While both implants revealed similar bone-to-implant contact, the nanoindentation demonstrated that the tissue quality was significantly enhanced around the HAp-coated implants, validating the postulated hypothesis. Osteoprogenitor activity after 2 weeks and a significant up-regulation in alkaline phosphatase (ALP) and osteocalcin in the coated implants were observed after 4 weeks post implantation. The latter observation suggested progressive mineralisation of the bone around the implant. Furthermore, adenosine triphosphatase, an osteoclast marker, was significantly higher for the treated implants, indicating gradual resorption of the calcium phosphate. Later, Roy *et al.*[15] also suggested that early bone tissue–implant integration was the result of nanostructured calcium phosphate coating that was plasma-coated. Using an *in vivo* rat distal femur, the formation of osteoid was observed on the coated implant surface.

In addition to the effects of surface morphology and surface chemistry, a study has also suggested that calcium phosphate nanocoatings and microcoatings of various crystallinities revealed different dissolution and re-precipitation properties that may increase early bone formation and bonding.[16] This was later supported by a study carried out by Surmeneva *et al.*[17] aimed at investigating the effect of the radio-frequency (RF) magnetron sputter deposited nano-hydroxyapatite thin films on *in vitro* cellular behaviour. Nanostructured, low-crystalline calcium phosphate coatings with thicknesses between 170 nm and 440 nm were deposited onto previously etched titanium substrates. The poorly crystalline thin calcium phosphate nanocoatings were observed to stimulate the attachment and proliferation of primary dental pulp stem cells and NCTC murine fibroblasts.

The production of thin and thick bioactive calcium phosphate films for clinical applications was anticipated during the last three decades using a number of conventional industrial coating techniques. Hench *et al.*[18,19] developed a unique technique that generates thick bioglass coatings with increased surface bioactivity.

An additional technique was developed in the early 1980s via plasma spray coating method to deposit calcium phosphate coatings (with thicknesses ranging from 30 $\mu$m to 2 mm) for bone ingrowth.[9,20]

The first pure HAp and tricalcium phosphate nanocoatings were developed in 1989. This approach involves dipping in alkoxide-based sol–gel-derived HAp solution to produce strong single or multilayered nanocoatings with thicknesses ranging between 70 nm and 100 nm.[1–3] Additional nanocrystalline calcium phosphate coating methods were introduced at a later stage, such as chemical and physical vapour deposition, electrophoretic deposition, RF magnetron sputtering, and biomimetic coatings.[21] While coatings based on self-assembly, nanocoatings, and bioglass coatings are effective, plasma spraying is the only industrial coating technique that has been widely applied commercially to dental and orthopaedic implants.

Plasma spraying uses an electrical discharge from a direct current arc or other sources such as RF to convert a carrier gas (such as argon) into gas plasma. The expansion of rapid gas induces speeds of up to 800 m/s. The powder is heated by the plasma into a partially liquid form and it is propelled towards the substrate. Plasma spraying is capable of depositing coating with thicknesses ranging from a few microns to a few millimetres.[22] Thermal spraying is a term that covers coating techniques such as detonation gun-spraying, flame-spraying, high-velocity oxygen flame-spraying, and plasma- spraying.

RF magnetron sputtering uses plasma produced by an RF field between the substrate holder and the sputtering target. In magnetron sputtering, a magnetic configuration is present below the target trapping electrons close to the target.[17] Ionisation in the sputtering gas is the result of the trapped electrons. A drop in voltage causes fraction of the ions to speed up towards the target. This is also referred to the plasma sheath and is present between the target and the plasma. The expulsion of secondary electrons and target species is due to the bombarding ions with the secondary electrons maintaining the discharge. The thickness of the thin film or coating is derived from the atoms that travelled from the target and bonded with the substrate material.

Chemical vapour deposition technique involves the chemical reactions of gaseous reactants on or near the vicinity of a heated substrate surface in order to produce coatings and films and it can be employed to produce single layer, multilayer, composite, nanostructured, and functionally graded coating materials at low processing temperatures with unique structure and well-controlled dimension. This atomistic deposition method can offer highly pure materials with structural control at atomic or nanoscale levels in addition to coating of complex-shaped biomedical prostheses and the fabrication of nanodevices and composites.[1,23]

Physical vapour deposition (PVD) involves the creation of vapour phase species either by means of evaporation, ion beam, laser ablation, or sputtering. PVD is a flexible synthesis method and is able for depositing, at the atomic or nanometre scale, thin-film materials with structural control. This control can be accomplished by carefully controlling the processing conditions. In most PVD-based processing approaches, it is not possible to fabricate a uniform coating on non-planar substrates without the use of multiple, spatially distributed sources or sophisticated substrate translation/rotation. This is due to the fact that the vapour atoms are produced in high vacuum which results in nearly collisionless vapour transport to the substrate. Therefore, only regions in the line-of-sight of the vapour source are coated.[24]

Pulsed laser deposition, more commonly known as 3D printing technology is a versatile and simple deposition technique and had been successfully applied for the fabrication of good-quality HAp thin films during the last decade. Due to its ability to generate well adhered and crystalline films and capable of restoring complex stoichiometries, this method had been proved to be more efficient and reliable and it will be used with increased frequency for industrial applications in the future.

The sol–gel technique is an attractive and versatile method that is also relatively easy to perform. It can be used to coat objects with complex shapes and can be utilised to fabricate ceramic coatings such as calcium phosphate from solutions by chemical means. The sol–gel technique has also been shown to improve both the osteointegration and mechanical properties as a result of the nanocrystalline grain structure produced.[25,26]

The history of the sol–gel approach (mainly for oxide coatings) dates back to the beginning of chemistry. It was first discovered as an application technology in 1846 when Ebelmen[27,28] observed with the hydrolysis and polycondensation of tetraethylorthosilicate (also referred to as tetraethoxysilane). In 1939, the first sol–gel patent was published covering the preparation of $SiO_2$ and $TiO_2$ coatings.[28]

The potential of sol–gel was recognised in 1954 by Roy and Roy[29] for the production of high-purity glasses as they discovered using traditional ceramic processing techniques were not adequate. Consequently, the first concept on the application of sol–gel technology to produce homogeneous multicomponent glasses was born.

Although pioneering work on the production method for nanosized HAp powder was initiated earlier,[30] the synthesis and application of pure and defect-free calcium phosphate nanocoatings did not start until years later. The first synthesis and application was aimed to coat a ceramic/titanium alloy knee implant which was initiated in the early 1990s. The reason for this delay was due to the synthesis problems encountered during the early stages.[31–33] Publications at that time named the process "sol–gel-derived thin films or nanocrystalline coatings".

## 5.3 Comparison between Plasma-sprayed Coatings and Nanocoatings

As mentioned previously, plasma spraying is without a doubt the most common technique for the deposition of calcium phosphate micro- and macro-coatings and it has been used comprehensively on a commercial scale as the primary coating process in the manufacture of dental and orthopaedic medical implants. However, at one stage, the production of plasma-coated calcium phosphate implants was shortened by the FDA for dental applications in the early 1990s due to the concerns related to mechanical disintegration and dissolution of the splats. These issues were solved with compositional changes to the calcium phosphate materials by adding fluorine and appropriate control of the processing parameters.

Although all of the above-mentioned problems have been resolved, the major concern from a clinical perspective with plasma-sprayed calcium phosphate coatings in orthopaedics for the last 25 years has been the adhesion of the splats generated during plasma-spraying and the dissociation of HAp to calcium oxide and other phosphate phases during the processing stages which has faster solubility within the physiological environment. The integrity of the coating–substrate interface, distribution and size of the pores formed, residual stresses, and defect population are also important and influential for long-term clinical stability. Furthermore, the strength of adhesion can be influenced by a number of different factors, such as spray parameters, substrate preparation, and powder characteristics.[9,20,34,35] Techniques such as micro-tensile adhesion testing and push and pull-off methods have been used by researchers to quantitatively determine the adhesion or bond strength of sprayed micro- and macrocoatings.[24,36]

The adhesive bond strength of plasma coatings is dependent on the processing conditions and varies from 5 to 25 MPa. Based on the findings from a number of studies, the bond strength was found to increase with increasing plate power, whereas it decreased with an increase in the working distance. The maximum adhesive bond strength of 25 MPa was achieved for samples prepared at 28 kW plate power with a working distance of 90 mm. After pull-off test, the fracture surfaces were classified as adhesive (delamination at the coating–substrate interface) and cohesive (delamination within the coating) failure.[24]

At the moment for clinical applications, nearly all dental and most orthopaedic implants and devices are macro- or micro-textured and then plasma coated to improve both mechanical interlocking and chemical bonding with the bone[37] (Figure 5.1). Due to the chemistry and thickness of the resultant coating, bone growth and adhesion can be easily initiated.

In contrast, new-generation nanocoatings are within the 70–200-nm range (as single or multilayer coatings) and due to their

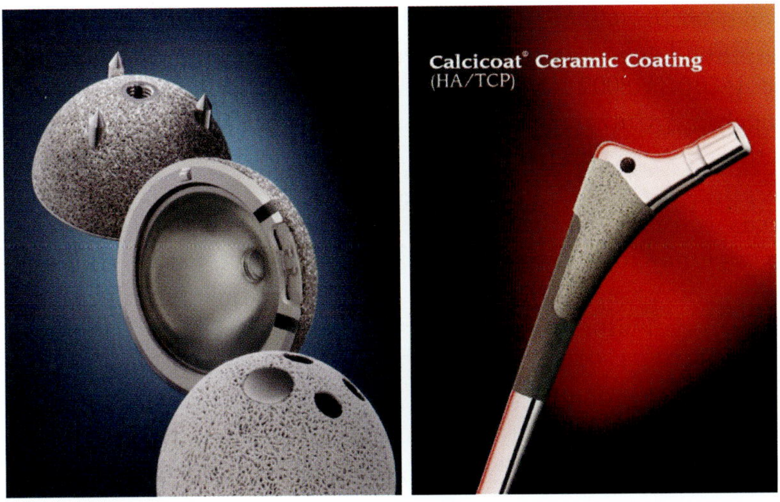


Figure 5.1 Micro-textured and HAp-coated acetabulum cups and femoral stem of hip surgery implants.

lack of micropores, they cannot generate bone mechanical interlock unless surfaces of the implants are micro- or macro-textured and then nanocoated. However, the biggest advantage of nanocoatings rest on their accelerated bioactivity due to their increased surface area owing to nanostructured grains (Figure 5.2) which promotes early osteointegration and faster healing.[1–4] Furthermore, they are produced at low temperatures with pure raw materials that the phases produced does not dissociate during production as it happens in the plasma coating technique. Application of these nanoscale coatings on macro- and micro-textured metallic surfaces generate stable mechanical interlock and chemical bonding due to their bone-like chemistry and structure.

## 5.4 Sol–Gel-derived Nanocoatings: Synthesis and Application Methods

While a number of coating production techniques are currently being applied to metallic implants on a research level, a wet chemical approach based on sol–gel in comparison to other production techniques is unique as it can be easily applied to complex shapes

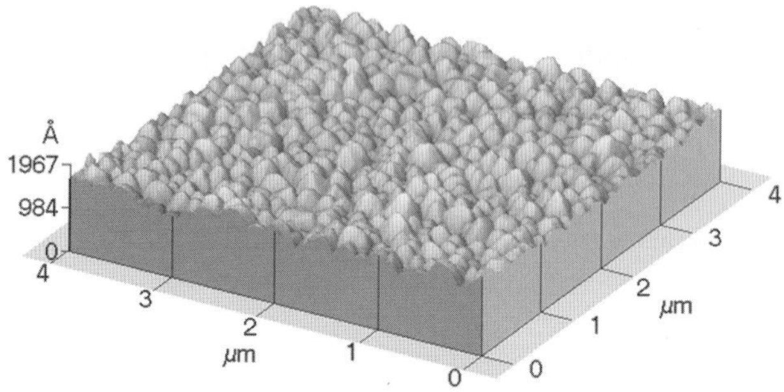


Figure 5.2 Atomic force microscopy (AFM) image of nano-HAp coating.

via dip coating or simply by a brush or spraying. Moreover, it permits the use of different chemical routes and their production and costs of precursor materials are relatively insignificant as only small amounts of materials are used. This technique also offers the advantage over other thin-film depositions in their thickness control, reduced pore size, and volume, and the surface area can be easily controlled through synthesis or production methods and grain size control.

A sol by definition is a suspension of colloidal particles in a liquid. The difference between a sol and a solution is that a solution is a single-phase system, whereas a sol may contain a two or more phases, solid–liquid system (colloidal system). Gravitational forces on these colloidal particles are negligible, and interactions are dominated by surface charges and short-range forces such as van der Waals forces. In certain circumstances, additional surfactants are used to assist dispersion and reduce or stop agglomeration.

The stability of the sol particles can be improved by reducing the surface charges. If the surface charge is significantly reduced, then gelation is triggered and the resultant product is capable of maintaining its shape without the assistance of a mould. Gels are considered a composite as they are composed of a solid network or skeleton that encloses a liquid phase or excess of solvent. They can be soft and have a low Young's modulus depending on their chemistry. This can be achieved through controlling the polymerisation of the hydrolysed starting compound.[38] A three-dimensional (3D) network forms

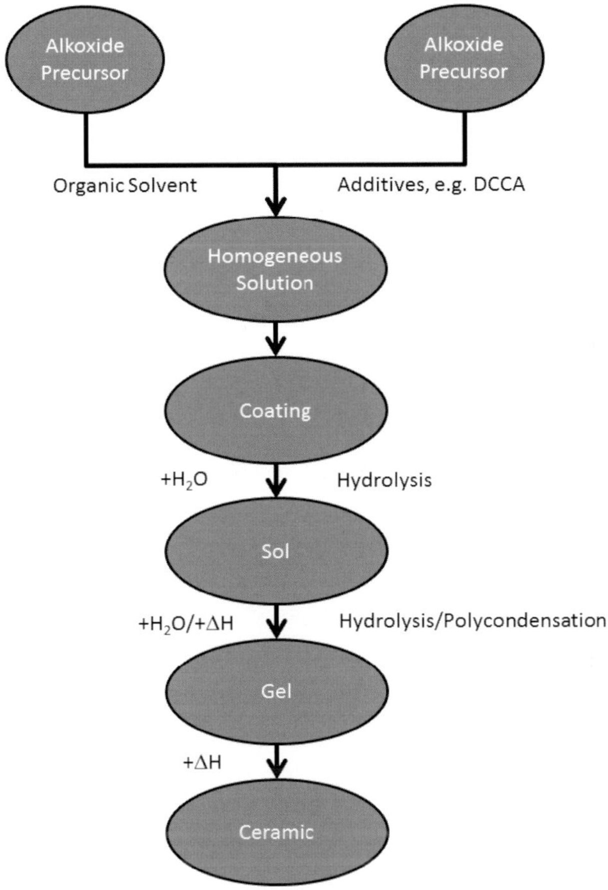


Figure 5.3 Diagram showing the production of a thin-film ceramic coating via the sol–gel technique.

in this case resulting ultimately in a high-molecular-weight polymeric gel. The resultant gel can be considered as a macroscopic arrangement of molecules that extends throughout the solution. The gelation point is referred to as the time taken for the formation of the last bond in this network. This gelation, depending on the process applied, can be used to produce powders, nanostructured monoliths, or nanocoatings (Figure 5.3).

The sol–gel technique can be classified as aqueous- or alcohol-based. Aqueous-based systems are generally carried out in the presence of water, whereas alcohol-based systems exclude water build-up until the hydrolysis stage. On the other hand, there are also non-hydrolytic sol–gel processes which do not require the presence of solvents at all. There are a number of excellent books, reviews, and past and current papers that summarises the synthesis, production methods, their characterisation, and testing as well as their applications can be found in Refs. 39–47.

Similarly, sol–gel precursors can be categorised as either alkoxides or non-alkoxides. Alkoxides are the preferred precursors for sol–gel production due to their easy volatility, which sometimes generate synthesis problems. Other compounds such as metal salts can also be used. This is often the case for Group I and II elements whose alkoxides are non-volatile, solid, and in many cases have a low solubility.[2–4]

The use of solvents is needed in the preparation of sol solutions, which are usually organic alcohols. Even though the main purpose of using solvent is to dissolve solid precursors, they are also used to minimise the effect of concentration gradients and dilute liquid precursors. The particular solvent employed can influence factors such as crystallisation temperatures and particle morphology.[45]

Due to the large amounts of organic materials needed (in the alkoxide process), cracking during the production stages caused by factors such as fast-drying can be an important issue on the soundness of the final products especially for solid monolithic bulk or foam production. In the case of monoliths, it is common that serious shrinkage occurs during drying. As a result, defects such as microcracking or edge cracking can occur (Figure 5.4). Specifically, cracking in thicker areas such as the edges are often a consequence of a number of factors such as phase separation, inhomogeneities, and thermal mismatch with the substrate or various problems related to the drying process.

The next stage is hydrolysis, or commonly referred to as drying and firing of the gel. During drying, excess solvents are removed from the pore network. A significant amount of shrinkage takes place during the drying and firing processes. Firing, which is also

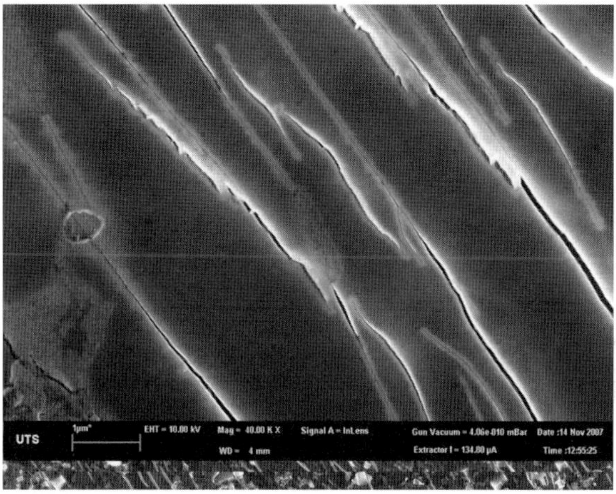


Figure 5.4 Scanning electron microscopy (SEM) image of edge microcracks within sol–gel-derived materials.

referred to as sintering, is the final stage in the production of ceramic materials via the sol–gel process. By heating dried gels to elevated temperatures, any remaining organic materials are combusted.

### 5.4.1 Calcium phosphate from alkoxide route

A number of sol–gel routes have been employed since the early 1990s for the production of synthetic calcium phosphate coatings. As previously mentioned, the first pure calcium phosphate nanocoatings were introduced by Ben-Nissan and coworkers in 1989 for a ceramic knee design[31–33] and it was based on the process of Masuda *et al.*[30] on HAp powder production. This method was modified to produce sound nanocoatings with highest purity and homogeneity. The coating technique was applied to a range of biomaterials such as 316L stainless steel, titanium alloys, cobalt chromium alloys, ceramics, glass, and a range of marine structure-derived calcium phosphates such as coral and foraminifera for bone graft and slow drug-delivery applications.[48]

The precursor solution was formed by mixing calcium diethoxide ($Ca(OEt)_2$) or calcium acetate monohydrate ($Ca(OAc)_2 \cdot H_2O$)

with ethylene glycol. A stoichiometric amount (Ca/P ration 1.67) of diethyl hydrogen-phosphonate $(C_2H_5O)_2P(O)H$ was added after the complete dissolution of calcium precursor. Coatings were formed using these solutions followed by subsequent heat treatment at low temperatures.[49,50] During the initial stages, the presence of calcium oxide in addition to HAp phase was detected if the solution was used without ageing.[49] Detailed study of this reactions, which occur at that temperature revealed a number of compounds that have been formed. The application of[31] PNMR technique, XRD, FTIR and DTA/TG, AFM, and SEM revealed that these compounds were mainly solutions of glycol mono- and diacetate, ethyl acetate (Figure 5.5). Most importantly, three major phosphorous-containing compounds have been identified. All of them were phosphonates having direct P–H bonds.[50–54]

In addition to using calcium alkoxide, other calcium precursors such as calcium acetate[55] and, in particular, calcium nitrate[56–59] have also been proved popular amongst researchers for the synthesis of calcium phosphate coatings on both the nano and microscales. A number of bioactivity studies have been conducted on these nanocoatings with thicknesses ranging from 20 to 1000 nm for a single-layered coating during the last decade.[55–59]

Studies have also been carried out to examine the effects of deposition parameters on the properties of the resultant coatings. It has been suggested that by changing the crystallinity and morphology of the coatings through different heat-treatment conditions can affect the attachment, proliferation, and differential behaviours of human osteosarcoma HOS TE85 cells.[56] The cells attached and proliferated well and expressed ALP and osteocalcin to a higher degree on the films with higher crystallinity as compared to the poorly crystallised film. Another study compared the difference between dip- and spin-coating techniques and the resultant nanocoatings.[55] Although questionable, their results suggested that compared to dip-coating technique, a more pure and more crystalline coating can be achieved using spin-coating technique. It should be noted that crystallinity depends on a number of other factors such as the raw material purity, synthesis chemistry,

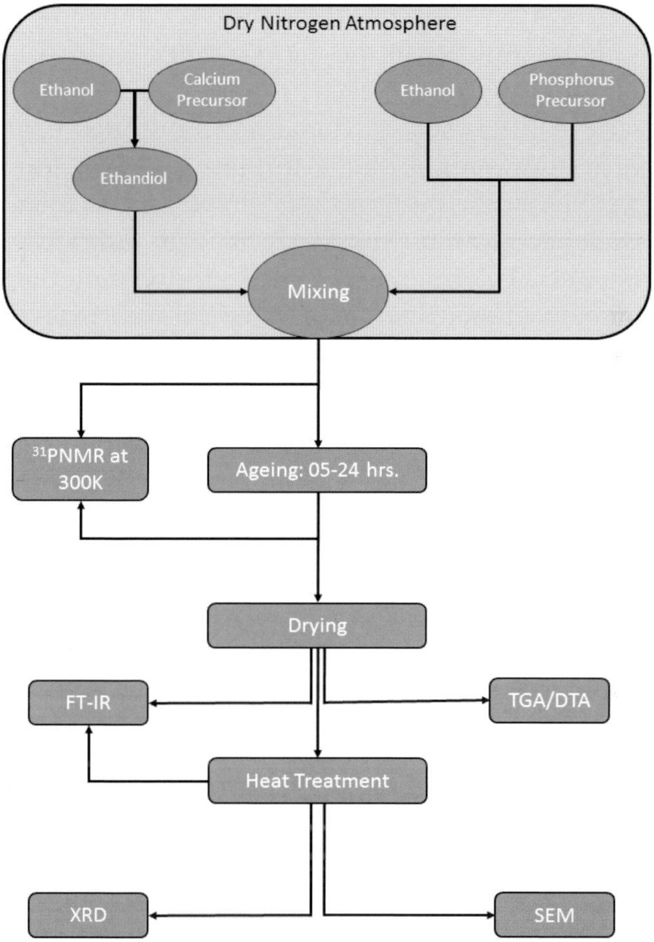


Figure 5.5 Procedure used for the synthesis of the calcium phosphate solutions, sample production, and analysis techniques. DTA: differential thermal analysis; FT-IR: Fourier transform infrared; NMR: nuclear magnetic resonance; SEM: scanning electron microscope; TGA: thermogravimetric analysis; XRD: X-ray diffraction.

temperature and the time of firing, heating and cooling rates, just to name a few.

Mudenda *et al.*[59] studied the microstructure evolution of calcium phosphate nanocoatings approximately 400-nm thick on wet and dry-etched micro-patterned titanium substrates. They observed the nanocoatings on wet-etched substrates contained larger calcium phosphate grains than the nanocoatings on dry-etched substrates. They suggested based on these observations that geometrically configured functionalised substrates could play a critical role at improving the bone–implant interface. Furthermore, they suggested that the patterned substrates would offer ideal calcium phosphate crystallinity and roughness to provide better osteointegration and anchorage of bone cells to the patterned implant surfaces. Nanocoatings due to their nanosized grainy structure supply this prerequisite and hence possibly it is one of the reasons for their early osteointegration.

As well reported, coating characteristics such as composition, crystalline features, and topography collectively impact the cell response. Kim *et al.*[60] investigated the dissolution behaviour of and *in vitro* cellular responses to a double-layered coating on a titanium substrate consisted of a calcium phosphate outer later and a fluoro-HAp inner layer with each layer approximately 600–800-nm thick. The *in vitro* solubility of the layered coating was between that of pure HAp and fluro-HAp film in a physiological solution and the dissolution profile was observed to be somewhat biphasic. The human osteoblast-like HOS TE83 cells cultured on the layered coating attached, spread, and grew favourably, and the proliferate rate was similar to that on pure HAp film and was significantly higher than that on the titanium substrate. The ALP activity and osteocalcin produced by the cells on the layered coating were also significantly higher than those on the titanium substrate.

Since the pioneering work of LeGeros in 1967 on minor substitutional elements in the HAp structure, researchers have examined the incorporation of different substitutional or trace elements into HAp in an effort to enhance the biological properties and performance of coated implants.[61–67] It has been observed through *in vitro* and *in vivo* studies that the addition of magnesium ions into the apatite crystals could contribute to the acceleration of osteoblast adhesion to apatite and promote bone formation.[68–70] To improve the biological properties, magnesium and fluorine ions using magnesium nitrate and hexafluorophosphoric acid as the Mg- and

F-precursors were incorporated into HAp simultaneously to form $FGMgCO_3Ap$–collagen composite.[66] The presence of fluorine was found to promote the incorporation of magnesium into the HAp crystal structure and the presence of magnesium-aided faster formation of a new apatite layer on the surface of the coating when submerged in simulated body fluid (SBF). High magnesium concentrations, on the other hand, promoted the formation of magnesium-substituted $\beta$-tri-calcium phosphate.

Another approach to enhance calcium phosphate-coated implants is through the addition of silicon into the apatite structure. It has been hypothesised that the presence of silicon could enhance the dissolution of HAp, resulting in an increase in the local concentrations of calcium and phosphate ions.[5,8] This in turn intensifies the amount of saturation in the medium and therefore enabling the precipitation of calcium phosphate phase.[62] This was later examined by Hijón *et al.*[63] in a study whereby silicon-substituted calcium phosphate nanocoatings with a thickness of 600 nm were deposited onto Ti–6Al–4V substrates. Using tetraethyl orthosilane as the silicon precursor, the coatings showed a homogeneous and crack-free surface composed of particles smaller than 20 nm. They observed that the apatite-forming ability on the surface of these silicon-substituted coatings was enhanced by the introduction of silicate into the HAp lattice and this layer was detected sooner after immersion in SBF.

## 5.5 Adhesion and Mechanical Properties of Nanocoatings

Understanding the long-term mechanical reliability of biomedical nanocoatings is pertinent in clinical applications. Depending on their applications, different orthopaedic or dental devices will require nano, micro-, or macrocoating materials with different designs, functions, and properties. Accurate measurement techniques are needed to determine the mechanical properties of these coatings as they can differ markedly from the bulk material. In particular, better and more reliable techniques are needed to quantitatively measure the adhesion strength and hardness of nanocoatings as well as the fracture toughness at the coating–substrate interface.[24] There have been significant advancements in the techniques and equipment capable of extracting the adhesion behaviour and mechanical properties of thin films particularly with the increasing trend towards deposition of nanometre-thick films on smaller devices. It is important to determine stresses and stability of nanocoatings as a function of deposition process used and the environment that they will be operated.

The stress developed in a film is made up of three components: the first component is intrinsic which is due to factors such as deposition, structure, and mode of growth of the crystallites; the second component is a result of the mismatch in thermal expansion between the film and the substrate; and the third component is related to externally applied loading due to the biomechanical functional loading.[71] Modelling studies has shown that the thermal expansion difference between the nanocoating and the substrate material becomes very critical in coatings thicker than 300 nm, while thinner coatings are relatively free from residual stresses.[72–74] Adhesion testing is essential to ensure that the nanocoatings will properly adhere to the substrate to which it is applied.

Many techniques have been used to measure the adhesion strength of coatings.[1,71,75] The most commonly used techniques for the characterisation of performance of nanocoatings on substrates can generally be divided into the measurement of adhesion strength and properties of applied coating. Some of the most commonly used techniques are nanoindentation, micro-tensile testing, scratch testing, adhesion and wear testing, pin-on-disk testing, pull-out test, and bending and bulge testing.[1,23,24,76]

Microtensile testing is an excellent method to determine the adhesion integrity of nanocoatings and it is ideal for determining the properties of both thick and thin coatings on a variety of ductile substrates and provides insights into the susceptibility of interfacial delamination through the application of controlled external stresses.[23,24]

In this adhesion test, the coating is deposited on a tensile coupon which can then be pulled in a universal testing machine or a specialised device that can be placed under the objective lens of an optical microscope or in an SEM with a special attachment. The cracking, damage evolution, and failure in the coating can be observed *in situ* or *ex situ* after application of specific strains. Tensile testing is advantageous in that the stress field is uniform along the gauge length of the sample and relatively small specimens can be used. The only prerequisite for this microtensile testing is that for analysis of the coating behaviour, and the residual stress, Young's modulus of the coating has to be obtained by other means such as from nanoindentation.[76–81] In a number of studies carried out by Roest *et al.*[76,82,83] using HAp nanocoatings on titanium alloys, it was observed that the adhesion of the nanocoatings was strong until the failure of the titanium alloy plates which showed slip bands (Figure 5.6) and the failure of the films demonstrated a tortuous path and branching demonstrating improved fracture toughness (Figure 5.7) and adhesion of the films.

The methodology and instrumentation involved with micro- and nanoindentation have been developed markedly since the

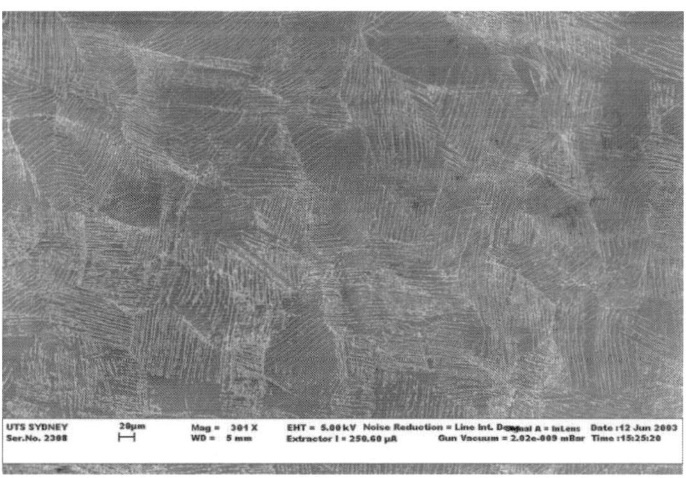


Figure 5.6 Slip lines and bands in HAp nanocoated titanium substrates at 300× magnification.

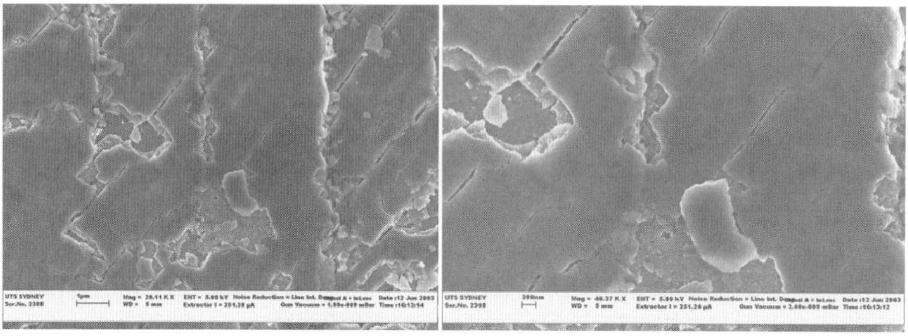

Figure 5.7 SEM of anodised and HAp-nanocoated commercially pure titanium sample after micro-adhesion testing at different magnifications (scale, 1 $\mu$m) (left) and (scale 200 nm) (right) showing crack branching and delamination.

early 1990s and it is now considered a routine and relatively effective means of obtaining accurate mechanical property measurements of nanocoatings such as Young's modulus and hardness. It is the method of choice by many researchers in the biomaterials field due to the simplicity of operation. The main requirements for obtaining the best possible results with such testing is dependent on variables such as the calibration of equipment, preparation of samples, and indenter tip shape.[76,84–86]

In the well-known conventional microindentation testing, a load is applied through a diamond tip indenter with a known geometry (usually Knoop, Rockwell, or Vickers) into the materials surface and then removed and the area of the residual impression is measured using optical means to obtain the material hardness, an example of which is shown in a study using both micro- and nanoindentation methods by Kealley *et al.*[87] on carbon nanotube-reinforced HAp composites. In nanoindentation testing, the size of the residual impressions is only often a few microns which make it extremely difficult to obtain a direct evaluation using optical techniques and instead electron microscopy is utilised.

All bioactive ceramic coatings such as calcium phosphate on dental and orthopaedic implants are subjected to a range of external biomechanical functional loading and resultant stresses which can cause delamination or perforation of the coating. In clinical applications, this can expose the substrate material to degradation and the released metal ions and/or coating particles under certain conditions may provoke a negative host response from the human tissues. For that reason, it is important that these medical coatings must retain its integrity when subjected to applied loads.

A number of approaches have been proposed to enhance the mechanical adhesion of sol–gel-derived calcium phosphate nanocoatings on titanium alloys. The application of a titanium oxide (titania; $TiO_2$)-anodised layer on titanium and its alloys before the deposition of calcium phosphate (usually HAp) is used to improve the bonding strength between the coating and the substrate.[76,82,83] These observations later were confirmed by Kim *et al.*[88] suggesting the improvement in bonding strength with the insertion of $TiO_2$ layer was due to its enhanced chemical affinity towards the calcium phosphate layer as well as towards the titanium substrate. They also reported that the bonding strength of the calcium phosphate/titania double-layer nanocoatings was markedly improved in comparison to that of single layer.

The formation of a titania-anodised layer can be achieved by a number of methods including appropriate anodising heat treatment or electrolysis. The development and subsequent characterisation including micro-adhesion behaviour of calcium phosphate nanocoatings on commercially pure titanium and Ti–6Al–4V alloy in relation to their anodising treatment was examined by Roest *et al.*[82] and Ben-Nissan *et al.*[89] The used substrate and the thickness of the anodised layer (as a function of applied voltage) were investigated and it was reported that they have the most profound influence on the interfacial adhesion of calcium phosphate. Furthermore, nanocoatings on Ti–6Al–4V exhibited superior interfacial bonding when compared to commercially pure titanium. It was suggested that this may be the result of the amounts of mixed anatase-rutile phases present and their phase transformations that improves chemical adhesion between the substrate and the nanocoated layer.[89] This was established using a number of characterisation techniques including secondary mass ion spectrometry method.[82,83]

The micro-adhesion testing of the titanium samples by Roest *et al.*[82,83] allowed for the calculation of the interfacial fracture energy and film toughness of the sol–gel coating on the anodised and un-anodised titanium samples. The HAp samples tested, although having a degree of variability in the results obtained, showed that the anodised layer improved the mechanical properties of the sol–gel nanocoating. The 50 V dilute acid-anodised sample had a fracture toughness value of 62.37 MPa $m^{1/2}$ compared to the non-anodised commercially pure titanium sample with a value of 37.82 MPa $m^{1/2}$, and a fracture energy of the film of 41.39 J $m^{-2}$ for the 50 V dilute sample and 15.22 J $m^{-2}$ for the non-anodised commercially pure titanium samples, showing improvement in both properties.

It is well established that modifications of substrate surfaces and heat-treatment temperatures used to produce coatings are the two major factors that influence the characteristics of the calcium phosphate coatings. An interesting and functional technique developed by Kokubo *et al.*[90] uses an alkali treatment to stabilise the coating and increase the adhesion strength. A study by Balakrishnan *et al.*[91] revealed that higher bond strength can be achieved on NaOH-treated Ti–6Al–4V alloy substrate surfaces. They suggested that the higher bonding strength can be attributed to factors such as the formation of a thin sodium titanium oxide layer and the high surface roughness.

Similarly, Park *et al.*[92] and Nagassa *et al.*[93] examined the use of hydrogen peroxide ($H_2O_2$) instead of NaOH to enhance the adhesion strength of the coating to the underlying substrate. Their data support the concept that surface topography or roughness and an oxide composition with appropriate thickness are significant factors that can influence the bond strength in both *in vitro* and *in vivo*. They postulated that modifications of titanium surfaces with $H_2O_2$ are more influential to protein adsorption. Additionally, it was hypothesised that treating titanium surfaces with $H_2O_2$ for 24 h, which induces micrometre-scale changes in roughness and increased protein adsorption, which enhances osteoblast attachment, mineralisation, and subsequent implant osteointegration.[93]

One of the most challenging approaches in the use of nanocoating has been the application of carbonate HAp nanolayer on natural structures such as coral (coralline structures) for bone graft and targeted slow drug delivery and implant coating applications. The patented method involves firstly converting the calcium carbonate coral structure to HAp by hydrothermal conversion and then coating it with nanocoating to improve the bioactivity and hence osteointegration.[25,94]

On the macroscale, coralline structures containing a range of 100–400-$\mu$m macropores can help vascularisation in clinical bone regeneration and bone fracture repair applications and it is similar to human trabecular structure in bulk morphology (Figure 5.8) but not in chemical composition ($CaCO_3$). Conversion of this calcium carbonate to HAp facilitates a better osteointegration to bone. On nano and meso-structure level, coralline HAp differs in morphology but contains interconnected meso- and nanopores that influence both biological and mechanical properties. By nanocoatings, these surfaces with extra carbonate HAp will enhance its bioactivity through the creation of a bioactive nanograin surface.[25,94–101]

In addition to the biological advantages, nanocoated coralline apatite also showed an increase in mechanical properties as a result

Figure 5.8 Coralline apatite bone grafts for bone regeneration. Courtesy of Biotomo Pty Ltd.

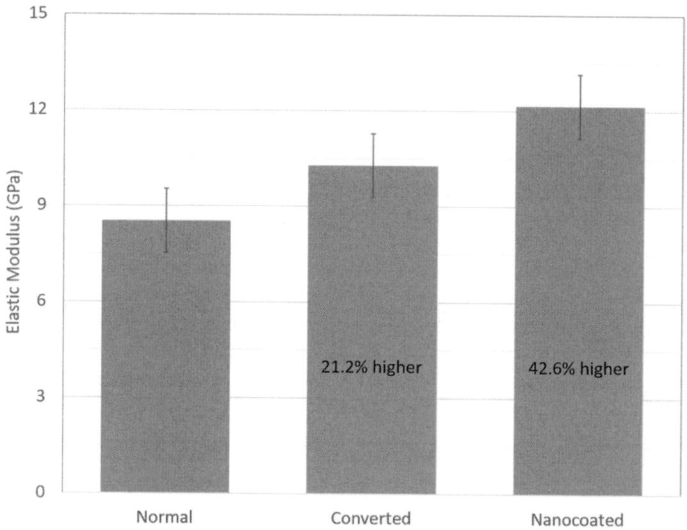


Figure 5.9 Variations in Young's modulus between normal coral, converted coral, and converted and nanocoated coralline apatite.

of the nanosized grain structure that covers the nano and mesopores. The increase in Young's modulus between normal coral to converted coral and nanocoated coralline apatite can be seen in Figure 5.9, which shows a 43% increase from natural coral to nanocoated apatite. Increase of approximately 450% in compression strength is also observed.

The original calcium carbonate structure of the coral contains a number of natural substitutions that include magnesium and strontium which is retained after conversion to calcium phosphates and improves osteointegration in clinical applications.[99–102] It was clearly shown that they accelerate osteogenic activity due to the combined influence of nanocoating in 3D scaffolds and structures.

### 5.5.1 Effects of surface topography

It was demonstrated that having nanocoating over meso- and nanoporous structures improves mechanical properties and performance due to the pore-filling effect of the nanocoating which penetrates to the nano and mesopores and that covering reduces the surface defects and improves properties of ceramic materials. Similarly, nanocoating and nanostructures on metallic substrates can have a number of advantages over polycrystalline large grain coated materials. Due to their extremely low grain size, nanocoated materials have large surface areas and hence increase in their reactivity is observed. This, on the other hand, introduces reduction in their sintering or densification temperatures and hence can be sintered at lower temperatures and at lower costs. Most importantly, the use of a nanolayer ceramic oxide can change the electrical or thermal insulating properties that are advantageous in many engineering and medical applications. For example, having a non-conductive oxide coating on medical orthopaedic or dental implant can stop or reduce the metal ion release which is a serious concern in clinical applications. Metal-ion release problems have increased since the introduction of the metal on metal bearings for the total hip replacement and hip resurfacing. On engineering, nanocoatings can help to introduce hydrophobic or hydrophilic properties and hence allow us to control their application in wear-resistant structures or self-cleaning areas. Nature, by design, has nanograins and nanostructures on external surfaces of many materials which improve performance. Natural structures use these extraordinary surface topographies in many different forms from colour control of butterfly wings due to light reflections to the unique filtering structures of the sea shells, corals, and foraminifera with their nanoporous interconnected open cell architectures. Surface morphologies and small crystalline grain structures of nanocoated materials can influence the biological properties with the acceleration of bonding to hard and soft tissues and early osteointegration of the implants in the clinical applications.

## 5.6 Biological Properties of Calcium Phosphate Nanocoatings

During the last several years, the use of zinc, magnesium, and strontium at certain doses has been shown to have beneficial effects in the stimulation of bone formation. Oliveira *et al.*[64] proposed that the local presence of strontium at the interface between an implant and bone could enhance osteointegration. Using an animal model, this effect on implant fixation in ovariectomised rats was examined by Li *et al.*[65] Strontium nitrate was used to replace 10 mol.% of calcium ions during the synthesis of calcium phosphate nanocoatings. When compared to pure calcium phosphate coatings, the strontium-substituted coatings showed similar phase composition, morphology, and surface roughness. Furthermore, after 12-week healing period, implants with the strontium-substituted coatings displayed improved osteointegration with an increase in bone area ratio and bone-to-implant compared to non-substituted calcium phosphate-coated implants.

Of particular interest is the gene expression profile obtained from magnesium-incorporated HAp nanocoatings at 21 days of cells growth. In this case, the attention was concentrated on the surface chemical and topographical effect since adipose-derived stem cells were directly seeded onto the specimen's surface. The results confirmed the gene expression tendencies observed at the previous cells growth, with a significantly higher expression of the osteopontin in the presence of magnesium. From these outcomes, it was proposed that the significance in the expression of osteopontin is likely to be due to the magnesium release rather than the surface topography, since the mesoporous surface topography and morphology was not altered by the addition of magnesium. They further suggested that osteopontin, which was a matrix molecule, strongly involved in the process of bone mineralisation, which is known to bind to HAp during crystal formation.[105]

The expression of osteopontin has been correlated with the differentiation and maturation stages of osteoprogenitors as well as osteoblasts.[105,106] Interestingly, osteopontin has been identified in the cement lines that demarcate the boundaries between old and new bone and therefore localised the bone formation front.[107] These evidences have been reported to show that osteopontin mediates osteoblast adhesion on the surfaces and bonds to the HAp crystals, during cell migration and attachment.[108] Thus, a higher expression of osteopontin in cells in contact with the magnesium-enriched mesoporous matrix may indicate that magnesium modulates the expression of this important protein.[109] Although one should not come to any conclusions with regard to the expression of the osteopontin, it may be one of the possible reasons why the same magnesium-modified mesoporous surface presented strong bone mechanical properties in a previously published animal study.[110] In brief, the removal torque values (RTQs) of mesoporous substrate surfaces with physically adsorbed magnesium ions were significantly higher compared to mesoporous surfaces not treated with magnesium after 3 weeks of studies.[110] These reverse transcription–polymerase chain reaction (RT-PCR) results are also consistent with pioneering investigations by Zreiqat *et al.*[101,102] which also showed a positive effect of magnesium-modified surfaces on osteoblast function.

Graf *et al.*[111] also investigated the time course of osteonectin expression in their pilot *in vitro* study, demonstrating that modification of a well-described anodised surface with magnesium or magnesium fluoride contributes to a faster organisation compared to commercially available anodised surfaces. Thus, they have shown that the gene expression levels at a longer time point clearly confirmed the osteogenic potential of the magnesium mesoporous coating to differentiate progenitor cells into the osteoblast phenotype. This is also ascertained by the results of ALP activity test since ALP represents a key enzyme secreted by osteoblast in the early stages of differentiation, which provides high concentration of phosphate at the sites of mineral depositions.[112]

Nanostructured calcium phosphate has been histologically proved to enhance osteointegration of implants; however, conventional techniques were not sufficiently sensitive to capture its biological effects fully. In a relatively recent dental conventional RTQ evaluation studies and gene expression in tissues around nanostructured calcium phosphate-coated implants, a number of interesting observations were made. In a rabbit model by using real-time RT-PCR, it was found that at 2 weeks, RTQ values and ALP expression were significantly higher and runt-related transcription factor 2 and tumour necrosis factor-$\alpha$ expressions were significantly lower in the coated than in the uncoated implants.[113] They reported that this indicated that inflammatory responses were suppressed and osteoprogenitor activity increased around the nanocoated surfaces. At 4 weeks, although RTQ values did not significantly differ between the coated and uncoated groups, ALP and osteocalcin were significantly up-regulated in the coated group, indicating progressive mineralisation of the bone around the implant. Moreover, an osteoclast marker, adenosine triphosphatase, which indicates acidification of the resorption lacunae, was significantly higher for the coated implants, suggesting gradual resorption of the calcium phosphate nanocoating. This study reveals detailed genetic responses to nanostructured calcium phosphate nanocoated implants and possibly confirms that the effect of surface modifications by nano and mesopores is significant during the osteointegration cascade.

In proof-of-concept studies of coralline apatite, we investigated the responses of immunoselected enriched human mesenchymal stem cells (eMSCs) to growth on transformed coral skeletons coated with HAp nanocoating. Human mesenchymal stem cells (MSCs; immuno-selected using STRO-1 antibody) were cultured *in vitro* on coated and uncoated Porites porite skeletons in basic media (10% FCS) for 14 days (Figure 5.10).

MSC proliferation and growth occurred throughout the skeletal architecture, bridging pores and filling voids. Significantly greater spread and density of MSCs were observed on sol–gel-coated skeletons compared to uncoated skeletons as shown by positive ALP staining and live/dead fluorescence staining.

At 14 days of ALP expression was significantly greater on the sol–gel-coated skeleton (0.005) compared to non-coated and plated eMSC controls and was directly equivalent to an osteogenic control. At 14 days, the quantity of eMSC osteocalcin and collagen using ELISA assays were significantly greater than tissue culture plastic controls in basic (un-supplemented) culture media.

MSC responses grown on transformed coated coral skeletons were significantly different to non-coated corals in their growth

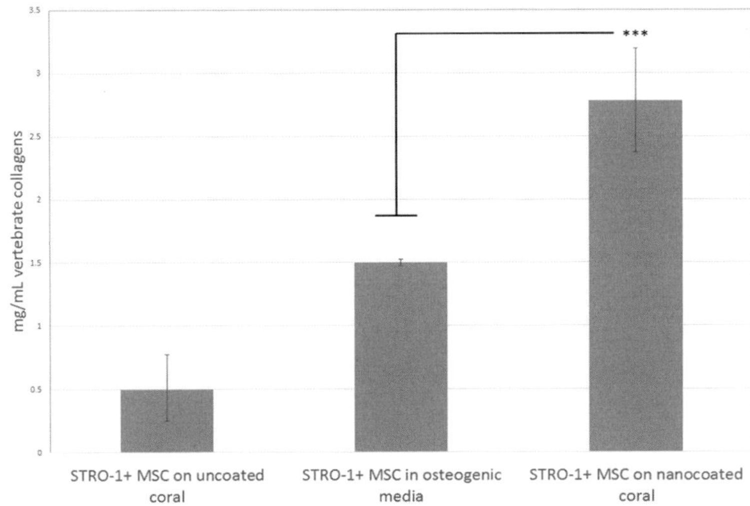


Figure 5.10 Collagen secretions by STRO-1 + selected hMSCs cultured on nanocoated and uncoated corals vs. osteogenic media.

and proliferation and up-regulation of bone cell-specific biochemicals such as ALP (early marker) and osteocalcin (late marker). These results show a natural bone-like behaviour of nanocoated coralline structures after MSC incorporation *in vitro*.

Further *in vivo* testing were carried out in sheep tibial and femoral component inserts and it was directed to evaluate the biological response of a coral-based converted bone graft implant in a head-to-head comparison with currently used bone graft implants produced from synthetic calcium phosphates. The study called for surgical implantation in a bilateral model applying implants in the distal femoral cancellous bone and proximal tibial cancellous bone. A bilateral femoral and tibial condyle model is proposed allowing two separate treatments to be placed in one individual. Time periods evaluated were 4, 8, 16, and 24 weeks. End-point analysis comprised linear bone (LBAR) apposition rate (fluorochromes) and light microscopy for biological response (Toluidene Blue). A total of 16 sheep were utilised with four sheep per time period. The photomicrographs clearly showed that the graft was behaving as both osteoconductive and surprisingly osteoinductive way, whilst maintaining an appropriate degree of structural integrity and strength. Results demonstrated that nanocoated modified coralline graft is clearly a controlled osseous-forming system.

## 5.7 Calcium Phosphate Nanolaminates and Nanocomposite Coatings

Multilayered, nanolaminated mixed nanocoatings can be considered nanocomposite coatings and can be described as a combination of two or more materials that includes a matrix nanocoated material mixed with nanoscale particles or laminated with different nanocoatings to produce required coated structures and properties. In the biomedical field, the matrix must be a biocompatible biogenic, polymeric, metallic, or ceramic material. With the composite approach, it is possible through the use of nanoparticles to manipulate mechanical properties of the composites such as strength closer to those of natural bone. These new-generation materials offer several attractive possibilities in the fields of tissue engineering for oral, maxillofacial, and orthopaedic applications.

It has long been established that porous bulk highly porous calcium phosphates cannot be used for load-bearing applications due to its unfavourable mechanical properties, but instead it can be used as a coating which utilises the ductility of the substrate materials.[1–3] This approach is currently being explored in the development of a new generation of nanocomposite coatings containing natural and synthetic nanomaterials such as carbon nanotubes, bioglass, collagen, chitosan, and silica to promote osteointegration and improve properties.[2] Addition of bone morphogenic proteins, growth factors, stem cells, and a number of pharmaceuticals are the current trends on multifunctional nanocoatings.

More recently, a study by Kelly *et al.*[114] suggested the addition of peptides will enable a titanium surface to be homogenous composite coated with HAp. The peptides used in their study consist of a combination of a HAp-binding peptide and a titanium-binding peptide. They also reported that the HAp–peptide composite coating showed no adverse effects on cell number or collagen deposition.

Synthetic polymers such as poly(lactic-co-glycolic acid) (PLGA) have also been used as secondary phases in nanocomposites with calcium phosphate. Negroiu *et al.*[115] suggested a nanocomposite structure consisted of calcium phosphate and polymeric material is expected to induce cellular responses at the interface with titanium implants. In their investigation, they observed cells grown on surfaces coated with a nanocomposite consisted of calcium phosphate and a maleic polyelectrolyte, sodium maleate–vinyl acetate copolymer, which induced a higher proliferation rate compared to calcium phosphate-coated surfaces, indicating the improvement in surface bioadhesion by the addition of a sodium maleate copolymer.

## 5.8 Concluding Remarks

Presently, a major disadvantage of metallic implants is their inability to adapt to the local tissue environment easily and bond strongly at the tissue–implant interfaces. Similar problems are observed in both dental and orthopaedic implants unless the surfaces modified to allow appropriate osteointegration. Cells do not adhere adequately or directly to most metallic surfaces even though implant materials such as cobalt chromium alloys and titanium and its alloys have been successfully utilised for nearly 50 years in orthopaedic surgery. The ideal mechanism for implant fixation is intimate tissue in-growth in both biological and physical means. The main purpose of using calcium phosphate nanocoatings is to alter the surface properties of implants to improve bioactivity and hence accelerate the healing process. This has been shown in calcium phosphate-coated implants with new bone formation in clinical use.

Nanocoatings, nanolaminates, and nanocomposite coatings in addition to other surface modification techniques are being developed and applied at the moment to create body-interactive materials in an attempt to promote tissue regeneration and repair. Surface coatings offer the possibility of modifying the properties of the medical implants and as a result, both the performance and reliability of the components are improved. Currently, a number of deposition techniques have been employed to fabricate micro- and nanoscale coatings of which some are bioactive, some wear protective, and/or corrosion resistant on different substrates.

During the next decade, it is envisaged that there will be an increase in the applications of biomedical implants and devices containing nanocoatings, nanolaminates, and nanocomposite coatings. These materials and devices will be used in implantable materials, stem cell and biogenic material-containing multi-functional coatings and scaffolds, biologically active membranes, slow drug-delivery systems, and bone grafts.

The physics and chemistry as well as the structural aspects of nanocoatings are different from the bulk materials we have been investigating for centuries. In terms of transformations at the nanoscale, structures and properties are also different. These issues are currently being investigated and generating interest in future application of knowledge gained. 3D scaffolding materials are one of the areas that will gain from this technology.

From the viewpoints of theoretical research, nanocoatings have become a vital tool aimed at understanding how the properties of biological and structural surfaces influence the interactions between implant and tissue at the nanoscale level to improve osteointegration and convert new-generation materials to behave in an osteoinductive manner. We can expect that surface modifications for the purpose of controlling tissue response as better understanding is achieved will create new avenues for the research and development of new and improved dental and biomedical implants and devices in a more systematic manner and at a faster rate than at present.

However, consideration has to be given to the potential risks associated with nanomaterials despite that the effects of nanotechnology are widely considered to be extremely beneficial. At present, there is a lack of regulatory standards governing the use of nanomaterials such as carbon nanotubes for dental and biomedical applications as well as no effective techniques to determine and assess long-term exposure risks to patients. All calcium phosphates are biodegradable and do not pose a long-term problem. On the other hand, all other non-biogenic nanomaterials should be considered carefully for their suitability in dental and biomedical applications. Problems or concerns may also arise from incomplete nanomaterial safety and toxicity profiles, which sequentially may influence a large range of long-term medical problems.

In nature, biomaterials are made with immaculate resource and energy efficiency using common, readily available substrates through self-assembly into highly organised hierarchies. Functional structures optimised to their environment are produced in this way. Studies of the way natural materials are constructed and the way they adapt to their environment will enable us to produce an exciting array of self-responsive structures and materials for regenerative medicine and structural applications. This will give us the opportunity to produce structures utilising nanocoatings and nanolaminated structures with intricate shapes and architectures that are tailored to their functions and do not break down. Biomimetic approaches can yield promising outcomes for application in the tissue engineering of skeletal tissues. One such approach involves configuring material environments at the molecular and macromolecular scales that try to mimic native extracellular matrix. The hope is to further extend this ongoing research towards the design of clinically relevant scaffolds for regenerative medicine using a unique set of self-organising hierarchical structures designed and synthesised according to biological principles of design. In the meantime, we are finally taking the initial steps towards this goal. We begin by harnessing as nature does with enormous precision and control inorganic molecules, nanoparticles, and nanocoatings to construct advanced functional biomaterials — some at the nanoscale level — with application in tissue engineering and pharmaceutical drug and gene delivery.

## Acknowledgements

The authors would like to acknowledge the continuous contributions made to this research area since the early 1990s by a number of students, postdocs, friends, and colleagues which without their help, guidance, and contribution, this task could not have been achieved. Just to name a few, we would like to thank Professors/Drs. Milev, Vago, Green. Kannangara, LeGeros, Swain, Roest, Latella, Bendavid, Heness, Chai, Evans, Gross, Hu, Russell, Pyke, Kohan, Conway, Tefry, Gilliott, Chou, Zreiqat, Valenzuela, Hanley, Gan, Macha, and Lewis.

## References

1. Ben-Nissan B, Choi AH, Green DW, Latella BA, Chou J, Bendavid A. Synthesis of hydroxyapatite nanocoatings by sol–gel method for clinical applications. In: Zhang S, editor. *Biological and Biomedical Coatings Handbook, Volume 1: Processing and Characterization*. London: CRC Press; 2011; pp. 37–79.
2. Choi AH, Ben-Nissan B, Matinlinna JP, Conway RC. Current perspectives: Calcium phosphate nanocoatings and nanocomposite coatings in dentistry. *J Dent Res* 2013;92:853–859.
3. Ben-Nissan B, Choi AH. Sol–gel production of bioactive nanocoatings for medical applications: part I: An introduction. *Nanomedicine* 2006;1:311–319.
4. Choi AH, Ben-Nissan B. Sol–gel production of bioactive nanocoatings for medical applications: part II: Current research and development. *Nanomedicine* 2007;2:51–61.
5. Hench LL. Bioceramics. *J Am Ceram Soc* 1998;81:1705–1728.
6. LeGeros R. Apatites in biological systems. *Prog Cryst Growth Charact Mater* 1981;41:1–45.
7. Hench LL. Bioceramics: From concept to clinic. *J Am Ceram Soc* 1991;74:1487–1510.
8. Hench LL, Wilson J. Surface active materials. *Science* 1984;226:630–636.
9. De Groot KR, Geesink R, Klein CPAT, Serekian P. Plasma sprayed coatings of hydroxylapatite. *J Biomed Mater Res* 1987;21:1375–1381.
10. LeGeros RZ. Properties of osteoconductive biomaterials: Calcium phosphates. *Clin Orthop Relat Res* 2002;395:81–98.

11. Hayakawa S, Kanaya T, Tsuru K, Shirosaki Y, Osaka A, Fujii E, Kawabata K, Gasqueres G, Bonhomme C, Babonneau F, Jäger C, Kleebe HJ. Heterogeneous structure and *in vitro* degradation behaviour of wet-chemically derived nanocrystalline silicon-containing hydroxyapatite particles. *Acta Biomater* 2013;9:4859–4867.
12. Albee FH, Morrison HF. Studies in bone growth. *Ann Surg* 1920;71:32.
13. Orsini G, Piattelli M, Scarano A, Petrone G, Kenealy J, Piattelli A, Caputi S. Randomized, controlled histologic and histomorphometric evaluation of implants with nanometer-scale calcium phosphate added to the dual acid-etched surface in the human posterior maxilla. *J Periodontol* 2007;78:209–218.
14. Jimbo R, Xue Y, Hayashi M, Schwartz-Filho HO, Andersson M, Mustafa K, Wennerberg A. Genetic responses to nanostructured calcium-phosphate-coated implants. *J Dent Res* 2011;90:1422–1427.
15. Roy M, Bandyopadhyay A, Bose S. Induction plasma sprayed nano hydroxyapatite coatings on titanium for orthopaedic and dental implants. *Surf Coat Technol* 2011;205:2785–2792.
16. Oh S, Tobin E, Yang Y, Carnes DL Jr, Ong JL. *In vivo* evaluation of hydroxyapatite coatings of different crystallinities. *Int J Oral Maxillofac Implants* 2005;20:726–731.
17. Surmeneva MA, Surmenev RA, Nikonova YA, Selezneva II, Ivanova AA, Putlyaev VI, Prymak O, Epple M. Fabrication, ultra-structure characterization and *in vitro* studies of RF magnetron sputter deposited nano-hydroxyapatite thin films for biomedical applications. *Appl Surf Sci* 2014;317:172–180.
18. Hench LL, Hench JW, Greenspan DC. Bioglass: A short history and bibliography. *J Aust Ceram Soc* 2004;40:1–43.
19. Hench LL, West JK. The sol–gel process. *Chem Rev* 1990;90:33–72.
20. Ducheyne P, Hench LL, Kagan A II, Martens M, Bursens A, Mulier JC. Effect of hydroxyapatite impregnation on skeletal bonding of porous coated implants. *J Biomed Mater Res* 1980;14:225–237.
21. Dorozhkin SV. Nanodimensional and nanocrystalline apatites and other calcium orthophosphates in biomedical engineering, biology and medicine. *Materials* 2009;2:1975–2045.
22. Gadow R, Killinger A, Stiegler N. Hydroxyapatite coatings for biomedical applications deposited by different thermal spray techniques. *Surf Coat Technol* 2010;205:1157–1164.
23. Ben-Nissan B, Latella BA, Bendavid A. Biomedical thin films: Mechanical properties. In: Ducheyne P, Healy KE, Hutmacher DW,

Grainger DW, Kirkpatrick CJ, editors. *Comprehensive Biomaterials*. Netherlands: Elsevier; 2011; pp. 63–73.

24. Ben-Nissan B, Choi AH, Bendavid A. Mechanical properties of inorganic biomedical thin films and their corresponding testing methods. *Surf Coat Technol* 2013;233:39–48.
25. Ben-Nissan B, Milev A, Vago R. Morphology of sol–gel derived nanocoated coralline hydroxyapatite. *Biomaterials* 2004;25:4971–4976.
26. Choi AH, Ben-Nissan B. Advancement of sol–gel technology and nanocoatings in Australia. *J Aust Ceram Soc* 2014;50:121–136.
27. Ebelmen J. Untersuchungen über die verbindung der borsaure und kieselsaure mit aether. *Ann Chim Phys Ser* 1849;57:319–355.
28. Geffcken W, Berger E. Änderung des reflexionsvermogens optischer gläser. German Patent 736411; 1939.
29. Roy DM, Roy R. An experimental study of the formation and properties of synthetic serpentines and related layer silicates. *Am Mineral* 1954;39:957–975.
30. Masuda Y, Matubara K, Sakka S. Synthesis of hydroxyapatite from metal alkoxides through sol–gel technique. *J Ceram Soc Jpn* 1990;98: 1255–1266.
31. Chai C, Ben-Nissan B, Pyke S, Evans L. Sol–gel derived hydroxyapatite coatings for biomedical applications. *Mater Manuf Process* 1995;10:205–216.
32. Ben-Nissan B, Chai C. Sol–gel derived bioactive hydroxyapatite coatings. In: Kossowsky R, Kossovsky N, editors. *Advances in Materials Science and Implant Orthopaedic Surgery*. NATO ASI Series, Series E: Applied Sciences; 1997, pp. 265–275.
33. Payten WM, Ben-Nissan B. The development of a modular ceramic knee prosthesis. In: Wise DL, Trantolo DJ, Lewandrowski KU, Yaszemski MJ, Gresser JD, Cattaneo MV, editors. *Biomaterials Engineering and Devices: Human Applications*. New York: Humana Press; 2000, pp. 309–335.
34. Ducheyne P, Radin S, Heughebaert M, Heughebaert JC. Effect of calcium phosphate ceramic coatings on porous titanium: Effect of structure and composition on electrophoretic deposition, vacuum sintering and *in vitro* dissolution. *Biomaterials* 1990;11:244–254.
35. Lin CK, Berndt CC. Measurement and analysis of adhesion strength for thermally sprayed coatings. *J Thermal Spray Technol* 1994;3:75–104.
36. Li H, Khor KA, Cheang P. Adhesive and bending failure of thermal sprayed hydroxyapatite coatings: Effect of nanostructures at

interface and crack propagation phenomenon during bending. *Eng Fract Mech* 2007;74:1894–1903.

37. Oonishi H, Yamamoto M, Ishimaru H, Tsuji E, Kushitani S, Aono M, Ukon Y. The effect of hydroxyapatite coating on bone growth into porous titanium implants. *J Bone Joint Surg Br* 1989;71:213–216.
38. Roy R. Ceramics by the solution-sol–gel route. *Science* 1987;238: 1664–1669.
39. Klein LC. *Sol–gel Technology for Thin Films, Fibers, Preforms, Electronics and Specialty Shapes*, 1st edition. New York: Noyes Publishing; 1989.
40. Scriven LE. Physics and application of dip coating and spin coating. *Mater Res Soc Symp Proc* 1988;121:717–729.
41. Brinker CJ, Clark DE. *Better Ceramics through Chemistry*, 3rd edition. Pittsburgh: Materials Research Society; 1988.
42. Brinker CJ, Scherer GW. *Sol–gel Science: The Physics and Chemistry of Sol–Gel Processing*, 1st edition. Boston: Academic Press; 1990.
43. Sakka S, Tanaka Y, Kokubo T. Hydrolysis and polycondensation of dimethyldiethoxysilane and methyltriethoxysilane as materials for the sol–gel process. *J Non-Cryst Solids* 1986;82:24–30.
44. Sakka S. Glasses and glass-ceramics from gels. *J Non-Cryst Solids* 1985;73:651.
45. Mazdiyasni KS. Powder synthesis from metal-organic precursors. *Ceram Int* 1982;8:42–56.
46. Sakka S, Kamiya K, Makita K, Yamamoto Y. Formation of sheets and coating films from alkoxide solutions. *J Non-Cryst Solids* 1984;63: 223–235.
47. Yoldas BE. Wide-spectrum anti-reflective coatings for fused silica and other glasses. *Appl Opt* 1984;23:1418.
48. Chou J, Valenzuela S, Green DW, Kohan L, Milthorpe B, Otsuka M, Ben-Nissan B. Antibiotic delivery potential of nano and micro porous marine structures derived β-TCP spheres for medical applications. *Nanomedicine* 2014;9:1131–1138.
49. Chai C, Gross K, Ben-Nissan B. Critical ageing of hydroxyapatite sol–gel solutions. *Biomaterials* 1998;16:2291–2296.
50. Milev A, Kannangara GSK, Ben-Nissan B. Morphological stability of plate-like carbonate hydroxyapatite. *Mater Lett* 2003;57:1960–1965.
51. Chai C, Gross KA, Kannangara K, Ben-Nissan B, Hanley L. Critical ageing and chemistry of nanocrystalline hydroxyapatite sol–gel solutions. *J Aust Ceram Soc* 1998;34:263–268.

52. Green DD, Kannangara GSK, Milev A, Ben-Nissan B. 31P solution state NMR investigation of the hydrolysis of a new alkoxide sol–gel hydroxyapatite. *Key Eng Mater* 2002;218–220:75–78.
53. Milev A, Kannangara GSK, Ben-Nissan B. Ligand substitution and complex formation in hydroxyapatite sol–gel system. *Key Eng Mater* 2002;218–220:79–84.
54. Chai CS, Ben-Nissan B. Bioactive nanocrystalline sol–gel hydroxyapatite coatings. *J Mater Sci Mater Med* 1999;10:465–469.
55. Malakauskaite-Petruleviciene M, Stankeviciute Z, Beganskiene A, Kareiva A. Sol–gel synthesis of calcium hydroxyapatite thin films on quartz substrate using dip-coating and spin-coating techniques. *J Sol–Gel Sci Technol* 2014;71:437–446.
56. Kim HW, Kim HE, Salih V, Knowles JC. Sol–gel modified titanium with hydroxyapatite thin films and effect on osteoblast-like cell responses. *J Biomed Mater Res A* 2005;74:294–305.
57. Carradò A, Viart N. Nanocrystalline spin coated sol–gel hydroxyapatite thin films on Ti substrate: Towards potential applications for implants. *Solid State Sci* 2010;12:1047–1050.
58. Choudhury P, Agrawal DC. Sol–gel derived hydroxyapatite coatings on titanium substrates. *Surf Coat Technol* 2011;206:360–365.
59. Mudenda S, Streib KL, Adams D, Mayer JW, Nemutudi R, Alford TL. Effect of substrate patterning on hydroxyapatite sol–gel thin film growth. *Thin Solid Films* 2011;519:5603–5608.
60. Kim HW, Knowles JC, Salih V, Kim HE. Hydroxyapatite and fluro-hydroxyapatite layered film on titanium processed by a sol–gel route for hard-tissue implants. *J Biomed Mater Res B Appl Biomater* 2004;71: 66–76.
61. Cai YL, Zhang S, Zeng XT, Wang YS, Qjan M, Weng WJ. Improvement on bioactivity with magnesium and fluorine ions incorporated hydroxyapatite coatings via sol–gel deposition on Ti6Al4V alloy. *Thin Solid Films* 2009;517:5347–5351.
62. Thian ES, Huang J, Best SM, Barber ZH, Bonfield W. Magnetron co-sputtered silicon-containing hydroxyapatite thin films — An *in vitro* study. *Biomaterials* 2005;26:2947–2956.
63. Hijón N, Victoria Cabañas M, Peña J, Vallet-Regí M. Dip coated silicon-substituted hydroxyapatite films. *Acta Biomater* 2006;2:567–574.
64. Oliveira AL, Reis RL, Li P. Strontium-substituted apatite coating grown on Ti6Al4V substrate through biomimetic synthesis. *J Biomed Mater Res B Appl Biomater* 2007;83:258–265.

65. Li Y, Li Q, Zhu S, Luo E, Li J, Feng G, Liao Y, Hu J. The effect of strontium-substituted hydroxyapatite coating on implant fixation in ovariectomized rats. *Biomaterials* 2010;31:9006–9014.
66. Yamasaki Y, Yoshida Y, Okazaki M, Shimazu A, Kubo T, Akagawa Y, Uchida T. Action of FGMgCO3Ap–collagen composite in promoting bone formation. *Biomaterials* 2003;24:4913–4920.
67. Kawamura H, Ito A, Miyakawa S, Layrolle P, Ojima K, Ichinose N, Tateishi T. Stimulatory effect of zinc-releasing calcium phosphate implant on bone formation in rabbit femora. *J Biomed Mater Res* 2000;50:184–190.
68. Relea P, Revilla M, Ripoll E, Arribas I, Villa LF, Rico H. Zinc, biochemical markers of nutrition, and type I osteoporosis. *Age Ageing* 1995;24:303–307.
69. Yamaguchi M, Oishi H, Suketa Y. Stimulatory effect of zinc on bone formation in tissue culture. *Biochem Pharmacol* 1987;36:4007–4012.
70. Chou J, Hao J, Kuroda S, Bishop D, Ben-Nissan B, Milthorpe B, Otsuka M. Bone regeneration of rat tibial defect by zinc-tricalcium phosphate (Zn-TCP) from porous foraminifera carbonate macrospheres. *Mar Drugs* 2013;11:5148–5158.
71. Lepienski CM, Pharr GM, Park YJ, Watkins TR, Misra A, Zhang X. Factors limiting the measurement of residual stresses in thin films by nanoindentation. *Thin Solid Films* 2004;447–448:251–257.
72. Gan L, Ben-Nissan B, Bendavid A. Modelling and finite element analysis of ultra-microhardness indentation of thin films. *Thin Solid Films* 1996;290–291:362–366.
73. Gan L, Ben-Nissan B. The effect of mechanical properties of thin films on nano-indentation data: Finite element analysis. *Comp Mater Sci* 1997;8:273–281.
74. Gan L, Ben-Nissan B. Finite element analysis and the behaviour of sol–gel derived thin films under ultra-microhardness indentation. *J Aust Ceram Soc* 1997;32:51–55.
75. Mittal KL. Adhesion measurement of films and coatings: A commentary. In: *Adhesion measurement of films and coatings*. Utrecht: VSP BV Publishers; 1995; p. 1.
76. Roest R, Latella BA, Heness G, Ben-Nissan B. Adhesion of sol–gel derived hydroxyapatite nanocoatings on anodised pure titanium and titanium (Ti6Al4V) alloy substrates. *Surf Coat Technol* 2011;205:3520–3529.

77. Oliver WC, Pharr GM. A new improved technique for determining hardness and elastic modulus using load and sensing indentation experiments. *J Mater Res* 1992;7:1564–1582.
78. Field JS, Swain MV. A simple predictive model for spherical indentation. *J Mater Res* 1993;8:297–306.
79. Swain MV, Mencik J. Mechanical property characterization of thin films using spherical tipped indenters. *Thin Solid Films* 1994;253:204–211.
80. Field JS, Swain MV. Determining the mechanical properties of small volumes of material from submicrometer spherical indenters. *J Mater Res* 1995;10:101–112.
81. Mencik J, Munz D, Quandt E, Weppelmann ER, Swain MV. Determination of elastic modulus of thin layers using nanoindentation. *J Mater Res* 1997;12:2475–2484.
82. Roest R. Physical and mechanical properties of hydroxyapatite and zirconia sol–gel coatings on anodised titanium substrate. M.Sc. Thesis. Sydney: University of Technology Sydney; 2004.
83. Roest R. Interfacial characterization of sol–gel derived coatings of hydroxyapatite and zirconia thin films with anodised titanium substrates. Ph.D. Thesis. Sydney: University of Technology Sydney; 2009.
84. Anast M, Bell J, Bell T, Ben-Nissan B. Precision ultra-microhardness measurements of sol–gel derived zirconia thin films. *J Mater Sci Lett* 1992;11:1483–1485.
85. Jamting AK, Bell J, Swain MV, Wielunski LS, Clissold R. Measurement of the micro mechanical properties of sol–gel $TiO_2$ films. *Thin Solid Films* 1998;332:189–194.
86. Saha R, Nix WD. Effects of the substrate on the determination of thin film mechanical properties by nanoindentation. *Acta Mater* 2002;50: 23–38.
87. Kealley CS, Latella BA, van Riessen A, Elcombe MM, Ben-Nissan B. Micro- and nano-indentation of a hydroxyapatite-carbon nanotube composite. *J Nanosci Nanotechnol* 2008;8:3936–3941.
88. Kim HW, Koh YH, Li LH, Lee S, Kim HE. Hydroxyapatite coating on titanium substrate with titania buffer layer processed by sol–gel method. *Biomaterials* 2004;25:2533–2538.
89. Ben-Nissan B, Choi AH, Roest R, Latella BA, Bendavid A. Adhesion of hydroxyapatite on titanium medical implants. In: Mucalo M, editor. *Hydroxyapatite (HAp) for Biomedical Applications*. Cambridge: Woodhead Publishing, Elsevier; 2015; pp. 21–51.

90. Kokubo T, Kushitani H, Ohtsuki C, Sakka S, Yamamuro T. Chemical reaction of bioactive glass and glass-ceramics with a simulated body fluid. *J Mater Sci Mater Med* 1992;3:79–83.
91. Balakrishnan A, Lee BC, Kim TN, Panigrahi BB. Hydroxyapatite coatings on NaOH treated Ti–6Al–4V alloy using sol–gel precursor. *Mater Sci Technol* 2007;23:1005–1007.
92. Park JH, Lee YK, Kim KM, Kim KN. Bioactive calcium phosphate coating prepared on $H_2O2$ treated titanium substrate by electrodeposition. *Surf Coat Technol* 2005;195:252–257.
93. Nagassa ME, Daw AE, Rowe WG, Carley A, Thomas DW, Moseley R. Optimisation of the hydrogen peroxide pre-treatment of titanium: Surface characterisation and protein adsorption. *Clin Oral Implants Res* 2008;19:1317–1326.
94. Ben-Nissan B. Natural bioceramics: From coral to bone and beyond. *Curr Opin Solid State Mater Sci* 2003;7:283–288.
95. Hu J, Russell JJ, Ben-Nissan B, Vago R. Production and analysis of hydroxyapatite from Australian corals via hydrothermal process. *J Mater Sci Lett* 2001;20:85–87.
96. Ben-Nissan B, Milev A, Green DD, Conway M, Vago R, Walsh W. Mechanical properties and characterisation of sol–gel coated coralline hydroxyapatite. *Key Eng Mater* 2002;218–220:379–382.
97. Ben-Nissan B, Russell JJ, Hu J, Milev A, Green DD, Vago R, Walsh W, Conway RM. Comparison of surface morphology in sol–gel treated coralline hydroxyapatite structures for implant purposes. *Key Eng Mater* 2000;192–195:959–962.
98. Lewis K, Valenzuela SM, Ben-Nissan B. Changes in the activity of osteoblast like cells with sol–gel derived hydroxyapatite and zirconia nanocoatings. *Key Eng Mater* 2008;361–363:633–636.
99. Ramaswamy Y, Ben-Nissan B, Roest R, Haynes D, Zreiqat H. Human osteoclasts behaviour on sol–gel derived carbonate hydroxyapatite coatings on anodized titanium alloy substrates. *Key Eng Mater* 2006;309–311:709–712.
100. Zreiqat H, Roest R, Valenzuela S, Milev A, Ben-Nissan B. Human bone derived cell (HBDC) behaviour of sol–gel derived carbonate hydroxyapatite coatings on titanium alloy substrates. *Key Eng Mater* 2005;284–286:541–544.
101. Zreiqat H, Valenzuela S, Ben-Nissan B, Roest R, Knabe C, Radlanski RJ, Renz H, Evans PJ. The effect of surface chemistry modification of titanium alloy on signalling pathways in human osteoblasts. *Biomaterials* 2005;26:7579–7586.

102. Chou J, Austin C, Doble P, Ben-Nissan B, Milthorpe B. Trace elemental imaging of coralline hydroxyapatite by laser-ablation inductively coupled plasma-mass spectroscopy. *J Tissue Eng Regen Med* 2014;9:515–520.
103. Lee LH. *Fundamentals of Adhesion*. New York: Plenum Press; 1991.
104. Kendall K. *Molecular Adhesion and Its Applications*. New York: Kluwer Academic/Plenum Publishers; 2001.
105. Merry K, Dodds R, Littlewood A, Gowen M. Expression of osteopontin messenger-RNA by osteoclasts and osteoblasts in modelling adult human bone. *J Cell Sci* 1993;104:1013–1020.
106. Lin Z, Rios HF, Volk SL, Sugai JV, Jin Q, Giannobile WV. Gene expression dynamics during bone healing and osseointegration. *J Periodontology* 2011;82:1007–1017.
107. Sodek J, Brunette DM, Feng J, Heersche JN, Limeback HF, Melcher AH, Ng B. Collagen-synthesis is a major component of protein-synthesis in periodontal ligament in various species. *Arch Oral Biol* 1977;22:647–653.
108. Katagiri YU, Sleeman J, Fujii H, Herrlich P, Hotta H, Tanaka K, Chikuma S, Yagita H, Okumura K, Murakami M, Saiki I, Chambers AF, Uede T. CD44 variants but not CD44s cooperate with beta 1-containing integrins to permit cells to bind to osteopontin independently of arginine–glycine–aspartic acid, thereby stimulating cell motility and chemotaxis. *Cancer Res* 1999;59:219–226.
109. Montezano AC, Zimmerman D, Yusuf H, Burger D, Chignalia AZ, Wadhera V, van Leeuwen FN, Touyz RM. Vascular smooth muscle cell differentiation to an osteogenic phenotype involves TRPM7 modulation by magnesium. *Hypertension* 2010;56:453–462.
110. Galli S, Naito Y, Karlsson J, He W, Andersson M, Wennerberg A, Jimbo R. Osteoconductive potential of mesoporous titania implant surfaces loaded with magnesium: An experimental study in the rabbit. *Clin Implant Dent Relat Res* 2014;doi: 10.1111/cid.12211.
111. Graf HL, Hofmann J, Tröger U, Schreckenbach J, Hilbig H. *In vitro* studies to screen for implant surface properties. *J Dent Implantol* 2009:250–261.
112. Zreiqat H, Howlett CR, Zannettino A, Evans P, Schulze-Tanzil G, Knabe C, Shakibaei M. Mechanisms of magnesium-stimulated adhesion of osteoblastic cells to commonly used orthopaedic implants. *J Biomed Mater Res* 2002;62:175–184.

113. Jimbo R, Coelho PG, Vandeweghe S, Schwartz-Filho HO, Hayashi M, Ono D, Andersson M, Wennerberg A. Histological and three-dimensional evaluation of osseointegration to nanostructured calcium phosphate-coated implants. *Acta Biomater* 2011;7:4229–4234.
114. Kelly M, Williams R, Aojula A, O'Neill J, Trzińscka Z, Grover L, Scott RA, Peacock AF, Logan A, Stamboulis A, de Cogan F. Peptide aptamers: novel coatings for orthopaedic implants. *Mater Sci Eng C* 2015;54:84–93.
115. Negroiu G, Piticescu RM, Chitanu GC, Mihailescu IN, Zdrentu L, Miroiu M. Biocompatibility evaluation of a novel hydroxyapatite-polymer coating for medical implants (*in vitro* tests). *J Mater Sci Mater Med* 2008;19:1537–1544.

# Chapter 6

# Bone Cements Utilised for the Reconstruction of Hard Tissue: Basic Understanding and Recent Topics

**Kanji Tsuru*, Yuki Sugiura† and Kunio Ishikawa‡**

*Faculty of Dental Science,*
*Kyushu University, Fukuoka, Japan*

* *tsuru@dent.kyushu-u.ac.jp*
† *ysugiura@dent.kyushu-u.ac.jp*
‡ *ishikawa@dent.kyushu-u.ac.jp*

**Abstract**

Various artificial bone substitutes such as hydroxyapatite, bone cement, and titanium or its alloys are used for reconstruction of hard tissues in the field of orthopaedic and oral maxillofacial surgeries. Among them, bone cement has paid much attention since it shows setting property that allows it to be injected by a syringe without incision. In the case of polymer-based cement consisting of acrylic polymer, it can be applied for fixation of metal implant or repair of vertebral fracture, which allows for prompt loading after surgery, whereas inorganic cements such as calcium phosphate, calcium sulphate, and calcium silicate are expected to bond directly to the surrounding tissues at the implanted site. This review focuses on acrylic polymer-based cement and calcium phosphate cements and also reveals a brief history, a composition, a setting mechanism, advantages and disadvantages, recent studies, and a future prospect of bone cements utilised for reconstruction of hard tissues.

## 6.1 Introduction

In the case of disease or trauma leading to physical loss of bone, a bone graft is required to reconstruct the bone tissue. Considering the patient's immune response, an autograft, using the patient's own bone,[1–3] is the first course of action since there is less risk of rejection and the bone itself has osteogenic and osteoconductive properties. Despite its safety and efficacy, this method suffers from several major drawbacks; additional surgery is required to harvest bone from a healthy site, causing trauma and pain, and this may result in surgical complications such as morbidity at the donor site. Furthermore, the quantity of available bone is limited. Of course, this method cannot be adopted in the case of large bone defects involving, for example, the loss of joint function. Therefore, artificial bone substitutes are necessary for patients needing bone tissue reconstruction.

Various artificial bone substitutes such as hydroxyapatite (HAp),[4] beta-tricalcium phosphate ($\beta$-TCP,[5,6] bone cements,[7–11] and titanium or its alloys[12,13] are used mainly for the reconstruction of hard tissue in the fields of orthopaedic and oral maxillofacial surgeries. Among these, bone cement has attracted much attention since it shows setting properties that allows injection using a syringe without the need for incision. The direct fitting of set cement to surrounding bone is also possible even if the bone defect is complicated in shape.

Cements used for the reconstruction of hard tissue are roughly classified into two types: polymer or inorganic. Acrylic polymer-based cements consisting of polymethyl-methacrylate (PMMA)[7,8] can be applied to metal implants or to the repair of vertebral fractures and allow prompt loading after surgery. Inorganic cement, represented by calcium phosphate cement,[9,10] displays osteoconductivity, i.e., the property of direct bonding to surrounding bone tissue at the implant site.

This review will focus on the basic understanding of and recent research into acrylic bone and calcium phosphate cements as representative of typical bone cements used for reconstruction of hard tissue.

## 6.2 Acrylic Bone Cement

### 6.2.1 Brief history and basic characteristics

The basic hardening reaction of acrylic bone cement, namely the synthesis of PMMA due to polymerisation of methylmethacrylate (MMA) monomers, was first reported by Röhm in the early 20th century.[7,8] Later, in 1943, commercially available acrylic cement that self-cured at room temperature was developed by Kulzer and Degussa.[7,8] This cement was first applied to the reconstruction of cranial bone defects and dental fillings. This was followed by use in the fixation of implants such as artificial joints and became popular in orthopaedics after the establishment of a surgical procedure for artificial hip joints using acrylic cement by Sir John Charnley in 1960.[14] To date, the typical composition consisting of an MMA liquid phase and a PMMA powder phase continues to be used without any major modification. Today, many kinds of acrylic bone cement are commercially available.[7,8,15,16]

The advantage of acrylic cement is that it shows superior mechanical properties similar to that of living bone. Table 6.1 summarises the mechanical properties of acrylic bone cement compared with natural bone.[15–20] Although the experimental conditions of each evaluation were different, the cement tensile strength was found to be 24–59 MPa[4,5,10] and the compressive strength is 73–117 MPa.[4,5,10] On the whole, the mechanical properties of acrylic bone cement are superior to that of other bone cements, including apatite cement, as shown in Section 3.1. This is why only acrylic bone cement can be used where large loads are applied such as the hip or knee joint and in balloon kyphoplasty, a minimally invasive procedure to repair spinal fractures. Although the high mechanical properties of this cement are an advantage, there are some

Table 6.1 Mechanical properties of acrylic bone cement compared with natural bone.

| | Tensile strength (MPa) | Compressive strength (MPa) | Young's modulus (Tensile) (GPa) | Fracture toughness $K_{IC}$ (MPa m$^{1/2}$) |
|---|---|---|---|---|
| Acrylic bone cement[4,5,9,10] | 24–59[4,5,10] | 73–117[4,5,10] | 16–41[4,5,10] | 0.5–3.7[9] |
| Cortical bone[10,11] | 50–150 | 100–230 | 7–30 | 2–12 |
| Cancellous bone[10,11] | 10–20 | 2–12 | 0.5–0.05 | |

disadvantages that affect clinical application, for example, this cement does not show osteoconductivity by itself despite being used directly in contact with living bone tissue.

### 6.2.2 Composition and polymerisation reactions

Acrylic bone cement consists of two components, a liquid and a solid phase.[7,8,14–16] A typical picture of commercial acrylic bone cement is shown in Figure 6.1. The main ingredient of the liquid phase is an MMA monomer. In the liquid phase, *N,N*-dimethyl-para-toluidine (DMPT), a tertiary amine, is incorporated as an activator, and hydroquinone is also included to inhibit the unintentional polymerisation of the MMA monomer in storage before use. The main ingredient of the solid phase is PMMA powder. In the solid phase, benzoyl peroxide (BPO) is included as an initiator. Almost all commercial products also contain barium sulphate ($BaSO_4$) or zirconium oxide ($ZrO_2$), which exhibit radiopacity.[15,16,21] Typical acrylic bone cement ingredients and their functions are summarised in Table 6.2.

MMA, with its reactive C=C double bond, participates in the polymerisation reaction.[7] As shown in Figure 6.2, first, a BPO molecule is activated by chemical reaction with DMPT to form a free radical called benzoyl oxiradical (upper). In the second step, this free radical reacts with the MMA monomer to start radical polymerisation (lower left). The free radical of the initiated chain reacts

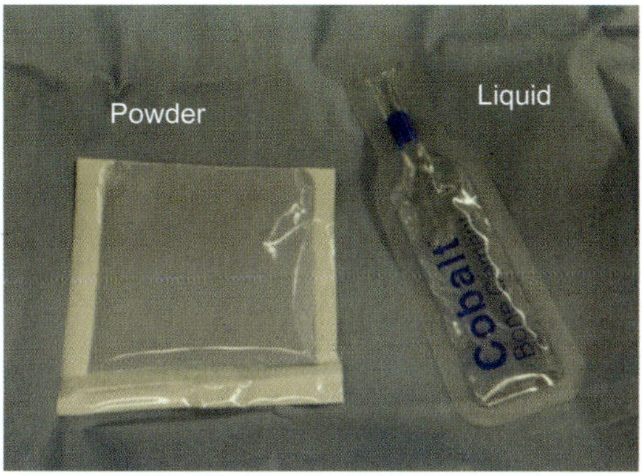


Figure 6.1 A commercially product of acrylic bone cement.

Table 6.2 Typical ingredients and their function in acrylic bone cement.

| Components | Ingredients | Role |
|---|---|---|
| Liquid phase | Methyl-methacrylate (MMA) | Monomer |
| | N,N-dimethyl para toluidine (DMPT) | Polymerization promoter |
| | Hydroquinone | Polymerization inhibitor |
| Powder phase | Polymethyl-methacrylate (PMMA) | Polymer |
| | Benzoyl peroxide (BPO) | polymerization initiator |
| | Barium sulfate ($BaSO_4$) or Zirconium oxide ($ZrO_2$) | Radiopacity |

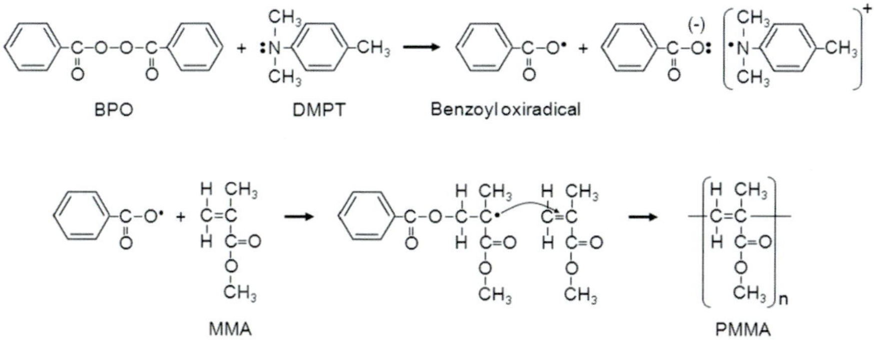


Figure 6.2 Radical polymerisation of MMA by activated BPO.

with other MMA monomers to grow a polymer chain (lower right). As a result, the molecular weight increases to become harder but less than $10^6$.[20,22]

The maximum temperature inside the curing cement reaches 60–80°C because the polymerisation of the MMA is an exothermic reaction (approximately 57 kJ [136 kcal] per mole of MMA); on the other hand, the real maximum temperature at the interface between bone and cement is approximately 40°C (48°C maximum) because of the heat dispersion due to thickness of the cement, blood circulation, and the thermally conductive metal stem.[7,8,23–25] However, no clinical problems along these lines have ever been reported. Nevertheless, clinicians need to understand that the temperature of curing cement increases and that these high temperatures may damage surrounding tissue due to necrosis.

In an actual application of the acrylic cement, the liquid and solid phases are mixed in a 2:1 [mass/mass] ratio and take 10–20 min to harden.[8,26] Use of PMMA powder contributes to the following: shortening the polymerisation time, thereby reducing the consequent heat generation, inhibition of volume shrinkage due to polymerisation, and reduction of the outflow of unreacted monomers into the bloodstream.

Figure 6.3 shows a time series of change in fluidity and elasticity from the start of the liquid and solid mixing. Problem arises when the timing of cement filling is too short or too long. When the timing is too short, unreacted MMA monomers outflowing into the bloodstream increase the risk of bone cement implantation syndrome[27] such as a decrease in blood pressure and cardiac arrest. When the timing is too long, the cement becomes incompatible with the bone or implant it is fit into. Therefore, to use this cement in clinical situations, one must be highly skilled.

### 6.2.3 Functional improvement

In clinical applications using acrylic bone cement, the occurrence of infection or long-term failure is reported even if the cement is applied properly.[28] These underlying problems are caused by

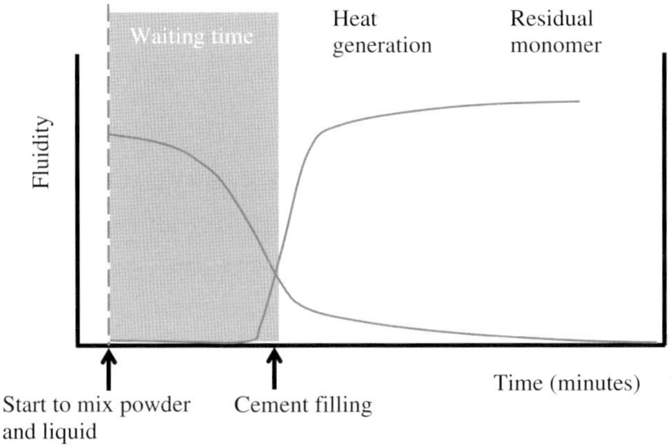


Figure 6.3 Time series of change in fluidity and elasticity from the start of liquid and solid mixing.

aseptic loosening due to indirect bonding between the implanted acrylic bone cement and the surrounding bone tissue.[29,30] This poor bone-bonding ability is inherent in the acrylic polymer.[30,31] As a consequence, improvement of inert acrylic bone cement is necessary to provide favourable tissue response.

#### 6.2.3.1 *Bonding between acrylic bone cement and host bone tissue*

Acrylic bone cement does not show bone-bonding ability even though it is used in direct contact with the host bone. Therefore, one of the main areas for the functional improvement of this cement is to achieve appropriate bonding to the host bone tissue. One of the initial attempts to improve the inert acrylic bone cement was by introducing a pore structure. This pore structure is useful in anchoring the inert material to the host tissues that penetrated into it.[32] To introduce a pore structure into acrylic bone cement, intrinsic forming agents,[33] water-soluble sucrose,[34] resorbable $\alpha$-tricalcium phosphate ($\alpha$-TCP),[34–37] or aqueous sodium carboxymethyl cellulose (CMC) gel[38–41] are employed and dispersed into the cement itself. In the case of $\alpha$-TCP, it not only creates a porous structure, but also shows biocompatibility and osteoconductivity.[42,43] The method using CMC gel is under continuous development, especially for clinical application in craniofacial surgery.[44,45]

Another attempt to improve inert acrylic bone cement is inclusion of bioactive ceramic powders such as HAp,[46,47] bioactive glass,[47] titanium oxide,[48,49] or $\beta$-TCP.[50,51] Although the inclusion of these powders does improve inert acrylic bone cement, a large amount of powder has to be used to achieve good bone-bonding ability; the ceramic powder needs to be exposed on the surface of hardened cement to ensure its osteoconductivity. However, large amounts of ceramic powder are difficult to handle and result in a low-viscoelasticity cement. This may promote high incidence of cement fracture. Regulation of ceramic particle size and shape is one of the strategies employed to solve this problem.[52] Combination with the other methods such as the pore inclusion described above is likely to be a better approach. Recently, biocompatible and porous acrylic bone cement was fabricated by the inclusion of both $\beta$-TCP particles and CMC gel and then used for the reconstruction of mandibular tissue.[53]

Another strategy to improve inert acrylic bone cement was reported by Ohtsuki and coworkers,[54–57] who determined that bioactive acrylic bone cement could be produced by the incorporation of alkoxysilane and calcium salts. This material design is based on the ability of apatite to form in the body. Apatite formation can be induced by a silanol group formed on the material as well as dissolution of calcium ions from the material. Inclusion of both alkoxysilane and calcium salts result in the formation of a silanol group and calcium ion release. Once apatite forms on the material, this can then directly bond to the host bone through the apatite layer. Histological analysis indicates that acrylic bone cement, with the appropriate amount of $\gamma$-methacryloxypropyl trimethoxysilane and calcium acetate added, exhibits osteoconductivity.[56]

#### 6.2.3.2 *Antibiotic acrylic bone cement*

When artificial material without any immune function is implanted, there is always a possibility of infection. Acrylic bone cement is no exception to this biophylactic function. Numerous studies regarding antibiotic-loaded acrylic bone cements have been reported since it was firstly introduced by Buchholz and Engelbrecht in 1970.[58] The rationale behind this is that antibiotics from the bone cement are gradually released, thereby delivering a local concentration able to inhibit most common pathogens.[59] This local concentration is much higher than that achieved by parenteral administration of the antibiotic.

As for the drugs used for antibiotic-loaded acrylic bone cement, numerous studies regarding use of gentamicin and tobramycin, known as aminoglycosides, have been reported.[60–64] The main criteria for selection are that the drugs have: water solubility due to their small molecular size, comparatively high thermal stability, low allergenicity, antibiotic activity at low concentrations, and exhibit a broad spectrum of antibiotic activity.[8,65–67] The use of vancomycin is also reported, but it was found that when loaded into bone cement, its antibiotic activity was lower than that of gentamicin using the same dosage.[68–70] Therefore, vancomycin is used together with gentamicin, tobramycin, or meropenem ($\beta$-lactam antibiotic) to obtain a wide spectrum of antibiotic activity.[59,71–73] In addition, this technique has been developed by the inclusion of porosity.[45] Briefly, colistin (peptide antibiotics) encapsulated with poly(lactic-co-glycolic acid) (PLGA) and CMC was mixed with PMMA cement components. This cement achieved 5-week continuous release of colistin owing to the PMMA/CMC/PLGA construct.[45]

In the past in Japan, when antibiotic-loaded acrylic bone cement was used for surgical operations, the medical doctors themselves had to mix the antibiotics with the powder phase of the acrylic bone cement. Nowadays, a gentamicin-loaded acrylic cement is produced by Biomet and has been commercially available since 2014.

Silver and chitosan are known to show antibacterial activity. Based on this idea, it has been reported that silver or chitosan nanoparticles have been dispersed into the matrix of acrylic bone cements.[74–76]

## 6.3 Calcium Phosphate Bone Cement

### 6.3.1 Brief history and basic characteristics

Apatite cement is a type of calcium phosphate cement that sets under bodily conditions. Initial findings regarding apatite cements were reported by Monma and Kanazawa in 1976 where α-TCP (α-$Ca_3(PO_4)_2$) set to form calcium-deficient hydroxyapatite (cd-HAp: $Ca_9(HPO_4)(PO_4)_5OH$).[77] Later, another apatite cement was produced by Brown and Chow in 1986.[78] This cement consisted of an equimolar mixture of tetracalcium phosphate (TTCP: $Ca_4(PO_4)_2O$) and dicalcium phosphate anhydrous (DCPA; $CaHPO_4$) powder that set to form HAp. In 1987, a different type of calcium phosphate cement, known as brushite-forming or dicalcium phosphate dihydrate (DCPD; $CaHPO_4 \cdot 2H_2O$) cement because it forms brushite under bodily conditions, was reported by Lemaitre *et al.*[79] This cement consists of a combination of β-TCP (β-$Ca_3(PO_4)_2$) and monocalcium phosphate monohydrate (MCPM; $Ca(H_2PO_4)_2 \cdot H_2O$) or phosphoric acid ($H_3PO_4$). The calcium phosphate ingredients used for calcium phosphate bone cements are summarised in Table 6.3.

Table 6.3 Calcium phosphate ingredients used for calcium phosphate bone cement.

| Calcium phosphates | Chemical formula | Ca/P ratio |
|---|---|---|
| Monocalcium phosphate monohydrate (MCPM) | $Ca(H_2PO_4)_2 \cdot H_2O$ | 0.5 |
| Monocalcium phosphate anhydrate (MCPA) | $Ca(H_2PO_4)_2$ | 0.5 |
| Dicalcium phosphate dihydrate (DCPD) | $CaHPO_4 \cdot 2H_2O$ | 1.0 |
| Dicalcium phosphate anhydrate (DCPA) | $CaHPO_4$ | 1.0 |
| α-tricalcium phosphate (α-TCP) | $Ca_3(PO_4)_2$ | 1.5 |
| β-tricalcium phosphate (β-TCP) | $Ca_3(PO_4)_2$ | 1.5 |
| Calcium-deficient Hydroxyapatite (cd-HAp) | $CA^{2+}{}_{10-x}(HPO_4)_x(PO_4)_{6-x}(OH)_{2-x}$ | <1.67 |
| Hydroxyapatite (HAp) | $Ca_{10}(PO_4)_6(OH)_2$ | 1.67 |
| Tetracalcium phosphate (TTCP) | $Ca_4(PO_4)_2O$ | 2.0 |

Today many kinds of calcium phosphate bone cements are commercially available.[9,80] Great feature of these cements is that they show not only setting properties, but also osteoconductivity. A bone defect can be filled with the cement without leaving a gap between the cement and surrounding bone.[9] This is the big advantage of calcium phosphate cement over bulk-type bone substitutes that need to be shaped to fit the bone defect, difficult when the defect is a complicated shape.

Table 6.4 summarises the mechanical properties of conventional calcium phosphate bone cements.[81–85] The diametral tensile strength (DTS) measurement is a useful method for determining the tensile strength of brittle materials such as calcium phosphate bone cements.[84] A schematic illustration of DTS measurements using a cylindrical specimen is shown in Figure 6.4. The specimen is placed with its principal axis perpendicular to the applied compressive load. External compressive loading generates tensile stress in the plane of the applied force as shown. The DTS value, $\sigma_t$, is determined by the following formula:

$$\sigma_t = \frac{2F}{\pi dt}, \tag{6.1}$$

where $F$ (N) is the fracture load in Newton, $d$ (mm) the diameter of specimen, and $t$ (mm) the thickness of the specimen. As a rough

Table 6.4 Mechanical properties of calcium phosphate cements.

| | CS (MPa) | DTS (MPa) | Young's modulus (Compressive) (GPa) | Fracture toughness $K_{IC}$ (MPa m$^{1/2}$) |
|---|---|---|---|---|
| α-TCP-based apatite cement | 40–102[84] | 9.0[85] | 3.7–8.5[84] | 0.4–0.5[84] |
| TTCP-DCPA-based apatite cement | 51–66[86] | 7.8–10.8[86] | | |
| Brushite cement | 7.4–51.0[88] | 1.4–5.9[88] | 25–80*[87] | 0.25–1*[87] |

*These values were obtained after being implanted into the defect of femoral condyle in rabbit for 2–16 weeks.

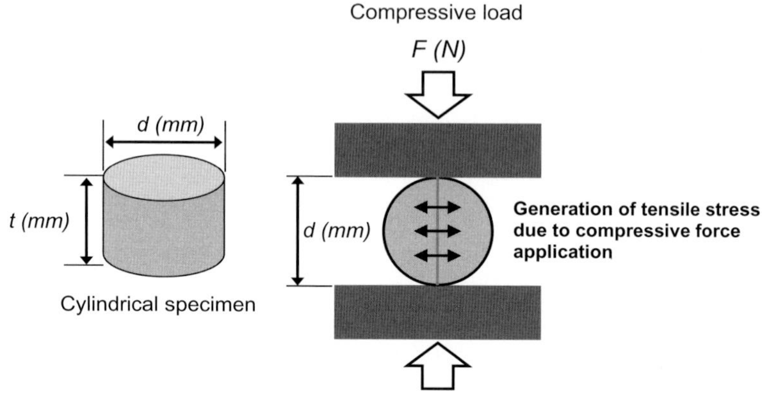


Figure 6.4 Schematic illustration of DTS measurements using a cylindrical specimen.

estimate, the compressive strength (CS) value is 5–6 times higher than the DTS value.

The mechanical properties of calcium phosphate cements are lower than those of living bone, as shown in Tables 6.1 and 6.4. Particularly, with regard to fracture toughness, $K_{IC}$, that of calcium phosphate bone cement is extremely low compared to acrylic bone cement or living bone. Therefore, calcium phosphate cement cannot be used for load-bearing repairs. When calcium phosphate cement is applied to a bone defect that experiences high loading, extra support with metallic biomaterials such as titanium or its alloys is necessary.

Among calcium phosphate cements, apatite cement shows superior mechanical properties to brushite cement. In one study, the mechanical properties of apatite cement increased after being implanted between femoral muscles in rats.[86] This increase can be ascribed to the faster conversion of the cement to HAp.[87] On the other hand, the mechanical properties of brushite cement do not increase in the early stages after implantation when compared to apatite cement.[88] This might be caused by the dissolution of DCPD. Interestingly, the compressive strength and the toughness of the cement–bone composite increased with increasing implantation periods and became higher when compared to those of normal bone at 12 weeks although those values decreased and returned to normal bone values at 16 weeks.[88] This indicates that brushite exhibits complete resorption and subsequent new bone formation after implantation.

### 6.3.2 Setting mechanism of calcium phosphate cement via the dissolution–precipitation reaction

The setting mechanism of calcium phosphate cement is similar to that of gypsum; basically, the process proceeds via a dissolution–precipitation reaction. Thermodynamically metastable calcium phosphates are used as the powder phase. This powder is then dissolved in a liquid, and precipitation of the thermodynamically stable phase then follows. Figure 6.5 shows the solubility of $Ca^{2+}$ in various calcium phosphates versus pH, calculated from the solubility product of each calcium phosphate.[89] As shown, HAp is the most thermodynamically stable phase in the pH range from neutral to alkaline, whereas brushite or DCPD becomes the most

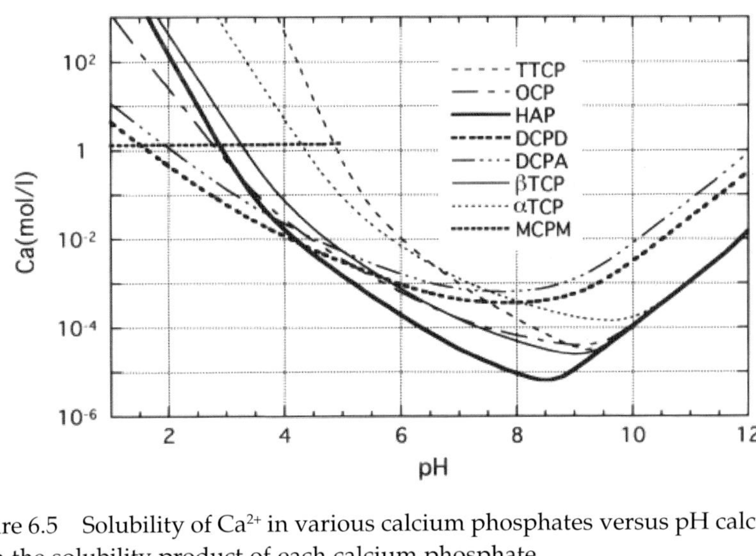


Figure 6.5 Solubility of $Ca^{2+}$ in various calcium phosphates versus pH calculated from the solubility product of each calcium phosphate.

thermodynamically stable phase when the pH is 4.2 or lower. In the case of apatite cement,[77,78] metastable phases such as α-TCP, TTCP, and DCPA are used as the powder phase. These powders dissolve in water to produce $Ca^{2+}$ and $PO_4^{3-}$ ions and thus reach supersaturation with respect to apatite; therefore, the most stable phase is precipitated via this dissolution–precipitation reaction. In the case of brushite or DCPD cement,[79,90,91] β-TCP is used as the powder phase in almost all cases. MCPM or $H_3PO_4$ gives acidic conditions so that DCPD, the most stable phase under acidic conditions as shown in Figure 6.5, is precipitated via the dissolution–precipitation reaction.

### 6.3.3 Setting reaction of apatite cement

#### 6.3.3.1 *Setting reaction of α-TCP-based apatite cement*[77]

The setting reaction of α-TCP-based apatite cement proceeds based on the dissolution–precipitation reactions shown in Eqs. (6.2) and (6.3).

**Dissolution:**

$$3Ca_3(PO_4)_2\ [\alpha\text{-TCP}] + 6H_2O \rightarrow 9Ca^{2+} + 6HPO_4^{2-} + 6OH^- \tag{6.2}$$

**Precipitation:**

$$9Ca^{2+} + 6HPO_4^{2-} + 6OH^- \rightarrow Ca_9(HPO_4)(PO_4)_5(OH)\ [\text{cd-HAp}] + 5H_2O \tag{6.3}$$

α-TCP powder is dissolved in water supplying $Ca^{2+}$ and $PO_4^{3-}$ ions and thus supersaturation is reached with respect to cd-HAp. As a result, cd-HAp crystals are precipitated and interlock with each other to set the cement. Figure 6.6 shows a typical scanning electron microscope (SEM) micrograph and transmission electron microscope (TEM) micrographs of a fractured surface of set α-TCP-based apatite cement. This cement results when a commercially obtained α-TCP powder (α-TCP-B, Taihei Chemical, Osaka, Japan) of approximately 5-μm diameter is mixed with distilled water at a liquid to powder (L/P) mixing ratio of 0.5 and stored in an incubator

Figure 6.6 Typical SEM and TEM micrographs of fractured surface of set α-TCP-based apatite cement.

maintained at 37°C for 24 h under 100% relative humidity. Cement assemblies with needle-like crystals can be observed in the TEM images. The needle-like crystals are the cd-HAp precipitates.

#### 6.3.3.2 *Setting reaction of TTCP-DCPA-based apatite cement*[78]

The setting reaction of TTCP-DCPA-based cement proceeds based on the dissolution–precipitation reactions shown in Eqs. (6.4)–(6.6).

**Dissolution:**

$$Ca_4(PO_4)_2O\ [TTCP] + H_2O \rightarrow 4Ca^{2+} + 2PO_4^{3-} + 2OH^- \qquad (6.4)$$

$$CaHPO_4\ [DCPA] \rightarrow Ca^{2+} + H^+ + PO_4^{3-} \qquad (6.5)$$

**Precipitation:**

$$2Ca_4(PO_4)_2O + CaHPO_4 \rightarrow 10Ca^{2+} + 6PO_4^{3-} + 2OH^- \rightarrow Ca_{10}(PO_4)_6(OH)_2 \text{ HAp} \quad (6.6)$$

TTCP–DCPA mixture powder is dissolved in water, supplying $Ca^{2+}$ and $PO_4^{3-}$ ions and thus supersaturation with respect to apatite is achieved. As a result, apatite crystals are precipitated and interlock with each other to set the cement. Figure 6.7 shows typical SEM and TEM images of a fractured surface of a set TTCP–DCPA-based cement. This was produced when an equimolar TTCP and DCPA powder mix was mixed with distilled water at an L/P of 0.3 and stored in an incubator maintained at 37°C for 24 h under 100%

Figure 6.7 Typical SEM and TEM micrographs of fractured surfaces of set TTCP-DCPA-based apatite cement.

relative humidity. Needle-like crystals interlock with each other to form cement assembles, though the size of the precipitated crystals is smaller than those observed in the $\alpha$-TCP-based cement.

### 6.3.4 Setting reaction of brushite cement[79,90,91]

The setting reaction of brushite cement proceeds based on the dissolution–precipitation reactions shown in Eqs. (6.7)–(6.11). For the reaction between $\beta$-TCP and MCPM in the presence of water, the equations are as follows[79,90]:

**Dissolution:**

$$Ca(H_2PO_4)\cdot H_2O\ [MCPM] \rightarrow Ca^{2+} + 2H^+ + 2HPO_4^{2-} + H_2O \quad (6.7)$$

$$Ca_3(PO_4)_2\ [\beta\text{-TCP}] + 2H^+ \rightarrow 3Ca^{2+} + 2HPO_4^{2-} \quad (6.8)$$

**Precipitation:**

$$Ca(H_2PO_4)\cdot H_2O + Ca_3(PO_4)_2 + 7H_2O \rightarrow 4Ca^{2+} + 4HPO_4^{2-} + 8H_2O \rightarrow 4CaHPO_4\cdot 2H_2O\ DCPD \quad (6.9)$$

Both $\beta$-TCP and MCPM dissolve in water to supply not only $Ca^{2+}$ and $PO_4^{3-}$ ions, but also $H^+$ ions and thus supersaturation with respect to brushite is reached under acidic conditions. As a result, brushite crystals are precipitated and interlock with each other to set the cement.

In the case of reaction between $\beta$-TCP and $H_3PO_4$, the equations are as follows[91]:

**Dissolution:**

$$H_3PO_4 \rightarrow 2H^+ + HPO_4^{2-} \quad (6.10)$$

**Precipitation:**

$$H_3PO_4 + Ca_3(PO_4)_2 + 6H_2O \rightarrow 3Ca^{2+} + 3HPO_4^{2-} \rightarrow 3CaHPO_4\cdot 2H_2O\ DCPD \quad (6.11)$$

$\beta$-TCP dissolves into phosphoric acid, supplying $Ca^{2+}$ and $PO_4^{3-}$ ions and thus reaches supersaturation with respect to brushite under acidic condition. As a result, brushite crystals are precipitated and interlock with each other to set the cement.

The setting time of brushite cement is very fast (approximately 30 s) compared to apatite cements because many $Ca^{2+}$ and $PO_4^{3-}$ ions are immediately released from calcium phosphate powders due to their higher solubility under acidic conditions. Under clinical conditions 30 s is too fast to handle the cement effectively. Therefore, the addition of a retarder to extend the setting time of brushite cement was studied; the details are described in Section 6.3.5.1.

### 6.3.5 Functional improvement of calcium phosphate cement

#### 6.3.5.1 *Setting time*

An appropriate setting time is required for the successful application of apatite cements. Apatite cements with longer setting times can cause inflammation, since unreacted powders are dispersed into the surrounding tissue, even though set apatite shows excellent tissue response.[92] Initially when developing TTCP–DCPA-based apatite cement, its setting time was 30–60 min, which is too long for clinical application. The setting time was improved by using a sodium hydrogen phosphate solution instead of water as the liquid phase.[87] For the apatite precipitation during the setting reaction, it was found that the dissolution of calcium phosphate powder or the supply of phosphate ions was the rate-determining reaction. This is the reason why phosphate solutions shorten the setting time of apatite cements.[87] This fast-setting cement showed higher mechanical strength than conventional apatite cement from the initial stages, without using sodium hydrogen phosphate solution.[86]

In the case of $\alpha$-TCP-based apatite cement, although the discovery that the setting property was due to the hydrate of $\alpha$-TCP resulted in a breakthrough in development, the setting time was still too long to be applied commercially. Therefore, trials were carried out to improve its setting properties. Addition of citric acid to the liquid phase was thought to improve the setting time of this cement since the initial setting was achieved by chelating the α-TCP and carboxyl groups from citric acid.[93,94] However, the addition of citric acid prolonged the phase transformation from α-TCP to cd-HAp.[93] Addition of apatite seed crystals, preferably those with low crystallinity, in the powder phase is also effective in shortening the setting time of apatite cement since it accelerates nucleation and crystal growth.[95–97] In fact, some commercially available apatite cements contain low-crystallinity apatite in its powder phase to shorten the setting time. Recently, the authors of this review proposed a new method of apatite seed crystal addition for α-TCP-based apatite cement, where it is partially hydrolysed to form cd-HAp on the surface of α-TCP powder.[98] The α-TCP powder was partially hydrolysed by immersion in distilled water for an appropriate period of time. As a result, low-crystallinity cd-HAp crystals could be introduced onto the surface of the α-TCP powder. Compared to the typical external seed addition method, cd-HAp seed crystals introduced by this method reside near the α-TCP powder. This location of seed crystals is favourable for the precipitation of new cd-HAp onto the seed cd-HAp and is followed by rapid growth, since the $Ca^{2+}$ and $HPO_4^{2-}$ ions are directly supplied from the surface of the α-TCP powder. This method is expected to be utilised for many commercial products of apatite cement to regulate not only the setting time, but also the mechanical properties.

In the case of brushite cement, the setting speed was too fast and, therefore, addition of a retarder was required. Calcium pyrophosphate and calcium sulphate were found to increase the setting time from 30 s to about 10 min.[99] Later, Bohner *et al*. reported that the effectiveness in increasing the setting time followed the order pyrophosphate > citrate > sulphate[100] and that their appropriate combinations can be used to control cement properties.[101]

### 6.3.5.2 *Anti-washout cement*

Apatite cement with anti-washout properties is required for clinical situations because unset cement may cause an inflammatory response, known as the crystalline inflammatory response, when unreacted powders disperse into soft tissue.[92] Apatite cement paste is hardened due to the interlocking of apatite crystals precipitated as a result of the dissolution–precipitation reaction described in Section 6.3.3. Therefore, the cement collapses due to washout when body fluids flow into the cement paste before the interlocking takes place. Although short setting time is one useful method of achieving anti-washout, the setting time is too short to manipulate the cement paste, and the side effects due to the addition of the accelerant must also be taken into account.

Basically, cement powders flow out when body fluids enter, so one effective method is to prevent the entry of body fluids into the cement paste before it sets. Based on this idea, anti-washout apatite cement can be produced by using a gelling agent that helps maintain bonding between the cement powders and also prevents body fluids entering the cement paste.[102–107] One of authors (K.I.) of this review recommended the use of a liquid phase containing 0.5 mass% sodium alginate. In this case, the total content of sodium alginate included in the cement paste becomes 0.125 mass% when the powder to liquid ratio (P/L [mass/ml]) is 4. This small amount of sodium alginate does not result in harm to the surrounding tissue.[102]

#### 6.3.5.3 *Porous calcium phosphate cement*

Construction of 3D macropore structures into artificial bone substitutes has proved popular in the development of commercial bone substitute products. These 3D structures, especially interconnected macropores of an appropriate size, can allow the penetration of cells, blood vessels, and tissues.[108–113] Since set calcium phosphate cement has no macropores, introduction of these into calcium phosphate bone cement has also attracted attention. Some studies on pore introduction to calcium phosphate cement have been reported.[114–124] Introduction of porogen or a pore-forming agent into calcium phosphate cements was attempted to fabricate pore structures in set cement. However, these methods resulted in only partially interconnected pores. Recently, in order to introduce interconnected pore structures into calcium phosphate cement, the authors of this review proposed a unique method using $\alpha$-TCP foam granules.[125] Briefly, $\alpha$-TCP foam was prepared by immersing polyurethane foam in a $\alpha$-TCP slurry such that the $\alpha$-TCP powder adheres to the surface of the polyurethane. This was then heated to burn off the polyurethane foam and sinter the $\alpha$-TCP powder.[126–128] After that, the $\alpha$-TCP foam was crushed and sieved to obtain an appropriate size of $\alpha$-TCP foam granules, retaining their porous structure. The obtained $\alpha$-TCP foam granules were used as the powder phase, and acidic calcium phosphate solution was used as the liquid phase, the same as the brushite-forming cement mentioned in Section 6.3.4. Since DCPD is the most stable phase at pH 2–4.5, when the $\alpha$-TCP foam granules and the acidic calcium phosphate solution were mixed, DCPD crystals precipitated on the surface of the $\alpha$-TCP foam granules. Thereby, the DCPD crystals were entangled with the bridge foam granules and each other. As a result, interconnected pores could be introduced into the set cement as shown in Figure 6.8. *In vivo* studies using experimental animals are waiting for interpretation on the effectiveness of the interconnected pores in this type of cement.

Figure 6.8 Typical SEM micrograph of interconnected pore-forming $\alpha$-tricalcium phosphate foam granule cement.

## 6.4 New concepts

### 6.4.1 Carbonate apatite cement

One of the advantages of apatite cement is its ability to be replaced by new bone. However, it has been reported that apatite cement is only minimally replaced by new bone.[129–135] The cause, hypothesised by the authors of this review, is that the replacement of apatite cement by new bone might be related to the carbonation of implanted apatite cement under bodily conditions.[136,137] The inorganic composition of bone is shown in Table 6.5.[138] Bone contains not only calcium phosphate, but also other elements. Among them, the concentration of carbonate ions is relatively high compared to the other trace elements where the content is less than 1 mass%. In fact, it has been reported that carbonate content shows a close relationship with osteoclastic resorption; carbonate-containing hydroxyapatite ($CO_3Ap$) shows a higher solubility in the weak acidic conditions in Howship's lacunae produced by the osteoclasts.[139] One of authors (K.I.) of this review reported that $CO_3Ap$

Table 6.5 Inorganic composition of adult human bone.[138]

| Chemical composition | Content (wt.%) |
|---|---|
| $Ca^{2+}$ | 38.4 |
| $PO_4^{3-}$ as P | 15.2 |
| Ca/P molar ratio | 1.71 |
| $Na^+$ | 0.9 |
| $Mg^{2+}$ | 0.72 |
| $K^+$ | 0.03 |
| $CO_3^{2-}$ | 7.4 |
| $F^-$ | 0.03 |
| $Cl^-$ | 0.13 |
| $P_2O_7^{4-}$ | 0.07 |
| Trace elements: $Sr^{2+}$, $Pb^{2+}$, $Zn^{2+}$, $Cu^{2+}$, $Fe^{3+}$, etc. | Balance |
| Total inorganic | 65.0 |
| Absorbed $H_2O$ | 10.0 |

granules fabricated by the dissolution–precipitation reaction showed osteoclastic bone resorption and new bone formation.[89] Based on these results, apatite cement showing adequate bone replacement can be produced if set calcium phosphate cement can transform into carbonate apatite after being implanted in the body. To achieve this, the starting materials must have carbonate ions as the constituting ions. For example, the case of a $CO_3Ap$-forming cement consisting of calcium carbonate ($CaCO_3$) and a possible candidate of DCPA; when $CaCO_3$ and DCPA are mixed with an aqueous solution, $CaCO_3$ and DCPA dissolve and supply $Ca^{2+}$, $CO_3^{2-}$, and $PO_4^{3-}$ ions. The solution is supersaturated with $CO_3Ap$ and is in its most thermodynamically stable phase. Precipitates of $CO_3Ap$ form, and then interlock with each other to set and harden. Therefore, a possible candidate for the starting material in the fabrication of $CO_3Ap$-forming cement is a combination of calcite $CaCO_3$ and DCPA. Although Takagi *et al.*[140] reported a combination of calcite ($CaCO_3$) and DCPA mixed with a $Na_2HPO_4$ solution, it failed to fully transform to apatite and the mechanical strength was low (approximately 1–1.5 MPa). Since the limited solubility of calcite might account for its failure to transform to apatite, the $CaCO_3$ polymorphs aragonite and vaterite, both with higher solubilities than calcite,[141,142] were used as the starting materials for apatite cement.[143,144] Recently, the authors of this paper succeeded in fabricating a bone cement that fully transforms into carbonate apatite using a combination of vaterite and DCPA mixed with sodium phosphate solutions.[144] The result of *in vivo* study based on this is awaited in order to confirm enhancement of osteoconductivity and resorbability for this type of $CO_3Ap$-forming cement.

### 6.4.2 Chelate-setting cement

Setting of typical apatite cement consisting of DCPA and TTCP proceeds based on the acid–base reaction. Thus, this setting reaction causes a pH change which results in inflammation to the surrounding tissue. Aizawa *et al.*[145–149] developed novel calcium-phosphate cements that set without this pH change. Briefly, the surface of calcium phosphate powder is modified by appropriate amounts of inositol phosphate (IP6), which is not only biocompatible, but also shows a chelating ability stronger than ethylenediamine tetraacetic acid. The IP6-modified calcium phosphate powder is simply mixed with water to be hardened by the chelating mechanism of IP6. It has been reported that α-TCP,[148] β-TCP,[149] commercial HAp,[145,146] or wet-synthesised HAp[147] has been used as a powder phase of calcium phosphate. This chelate-setting cement has also been studied for application to anti-bacterial activity[150] and anti-tumour action.[151,152]

## References

1. Moore WR, Graves SE, Bain GI. Synthetic bone graft substitutes. *ANZ J Surg* 2001;71:354–361.
2. Kurz LT, Garfin SR, Booth RE. Harvesting autogenous iliac bone grafts. A review of complications and techniques. *Spine* (*Phil Pa 1976*) 1989;14:1324–1331.
3. Fernyhough JC, Schimandle JJ, Weigel MC, Edwards CC, Levine AM. Chronic donor site pain complicating bone graft harvesting from the posterior iliac crest for spinal fusion. *Spine* (*Phil Pa 1976*) 1992;17: 1474–1480.
4. Aoki H. *Medical Applications of Hydroxyapatite*. Tokyo, St Louis: Ishiyaku EuroAmerica, Inc., 1994.
5. Galois L, Mainard D, Delagoutte JP. Beta-tricalcium phosphate ceramic as a bone substitute in orthopaedic surgery. *Int Orthop* 2002;26:109–115.
6. Horowitz RA, Mazor ZM, Foitzik C, Prasad H, Rohrer M, Palti A. β-Tricalcium phosphate as bone substitute material: Properties and clinical applications. *Int J Dent Implant Biomater* 2009;1(2):2–11.
7. Kuehn KD, Ege W, Gopp U. Acrylic bone cements: Composition and properties. *Orthop Clin N Am* 2005;36:17–28.
8. Magnan B, Bondi M, Maluta T, Samaila E, Schirru L, Dall'Oca C. Acrylic bone cement: Current concept review. *Musculokelet Surg* 2013;97:93–100.
9. Ishikawa K. Bioactive ceramics: Cements, in Ducheyne P, Heary KE, Hutmacher DW, Grainger DW, Kirkpatrick CJ, editor, *Comprehensive Biomaterials*, Vol. 1, San Diego, CA, USA: Elsevier, 2011; pp. 267–283.

10. Dorozkin SV. Calcium orthophosphate cements for biomedical application. *J Mater Sci* 2008;43(9):3028–3057.
11. Hughes E, Yanni T, Jamshidi P, Grover LM. Inorganic cements for biomedical application: Calcium phosphate, calcium sulphate and calcium silicate. *Adv Appl Ceram* 2015;114:65–76.
12. Brown SA, Lemons JE. *Medical Applications of Titanium and Its Alloys: The Material and Biological Issues*. Philadelphia, PA:ASTM;1996.
13. Rack HJ, Qazi JI. Titanium alloys for biomedical applications. *Mater Sci Eng C* 2006;26(8):1269–1277.
14. Charnley J. Anchorage of the femoral head prosthesis to the shaft of the femur. *J Bone Joint Surg* 1960;42B:28–30.
15. Lewis G. Properties of acrylic bone cement: State of the art review. *J Biomed Mater Res* 1997;38:155–182.
16. Lewis G. Alternative acrylic bone cement formulations for cemented arthroplasties: Present status, key issues, and future prospects. *J Biomed Mater Res B Appl Biomater* 2008;84:301–319.
17. Kuehn KD, Ege W, Gopp U. Acrylic bone cements: Mechanical and physical properties. *Orthop Clin N Am* 2005;36:29–39.
18. Saha S, Pal S. Mechanical properties of bone cement: A review. *J Biomed Mater Res* 1984;18:435–462.
19. Hench LL, Wilson J. Introduction, Chapter 1, in Hench LL, Wilson J, editor, *An Introduction to Bioceramics*, Vol. 1, Singapore: World Scientific; 1993; pp. 1–24.
20. Haas SS, Brauer GM, Dickson G. A characterization of polymethylmethacrylate bone cement. *J Bone Joint Surg Am* 1975;57(3):380–391.
21. Artola A, Goõi I, Gil J, Ginebra P, Manero JM, Gurruchaga M. A radiopaque polymeric matrix for acrylic bone cements. *J Biomed Mater Res B Appl Biomater* 2003;64:44–55.
22. Topoleski LD, Ducheyne P, Cuckler JM. Microstructural pathway of fracture in poly(methyl methacrylate) bone cement. *Biomaterials* 1993;14:1165–1172.
23. Biehl G, Harms J, Hanser U. Experimental research on heat-development in bone during polymerization of bone cement. *Arch Orthop Trauma Surg* 1974;78(1):62–69.
24. Reckling FW, Dillon WL. The bone–cement interface temperature during total joint replacement. *J Bone Joint Surg Am* 1977;59(1):80–82.
25. Webb JC, Spencer RF. The role of polymethylmethacrylate bone cement in modern orthopaedic surgery. *J Bone Joint Surg Br* 2007;89: 851–857.

26. Goodman S. Wear particulate and osteolysis. *Orthop Clin North Am* 2005;36:41–48.
27. Donaldson AJ, Thomson HE, Harper NJ, Kenny NW. Bone–cement implantation syndrome. *Br J Anaesth* 2009;102(1):12–22.
28. Berry DJ, Harmsen WS, Cabanela ME, Morrey BF. Twenty-five-year survivorship of two thousand consecutive primary Charnley total hip replacements: Factors affecting survivorship of acetabular and femoral components. *J Bone Joint Surg Am* 2002;84(2):171–177.
29. Savarino L, Stea S, Ciapetti G, Donati ME, Alvergna P, Pizzoferrato A. Microstructural investigation of bone–cement interface. *J Biomed Mater Res* 1995;29:701–705.
30. Hughes KF, Ries MD, Pruitt JA. Structural degradation of acrylic bone cements due to *in vivo* and simulated aging. *J Biomed Mater Res Part A* 2003;65(2):126–135.
31. Bettencourt A, Calado A, Amaral J, Alfaia A, Vale FM, Monteiro J, Montemor MF, Ferreira MGS, Castro M. Surface studies on acrylic bone cement. *Int J Pharm* 2004;278:181–186.
32. Hulbert SF, Morrison SJ, Klawitter JJ. Tissue reaction to three ceramics of porous and non-porous structure. *J Biomed Mater Res* 1972;6: 347–374.
33. Nathanson D, Gettleman L, Schnitman P, Shklar G. Histologic response to porous PMMA implant materials. *J Biomed Mater Res* 1978;12:13–33.
34. Rijke AM, Rieger MR. Porous acrylic cement. *J Biomed Mater Res* 1977;11:373–394.
35. Beruto DT, Mezzasalma SA, Capurro M, Botter R, Cirillo P. Use of α-tricalcium phosphate (TCP) as powders and as an aqueous dispersion to modify processing, microstructure, and mechanical properties of polymethylmethacrylate bone-substitute compounds. *J Biomed Mater Res* 2000;49:498–505.
36. Beruto DT, Botter R, Fini M. The effect of water in inorganic microsponges of calcium phosphates on the porosity and permeability of composites made with polymethylmethacrylate. *Biomaterials* 2002;23: 2509–2517.
37. Fini M, Giavaresi G, Aldini NN, Torricelli P, Botter R, Beruto D, Giardino R. A bone substitute compound of polymethylmethacrylate and α-tricalcium phosphate: Results in terms of osteoblast function and bone tissue formation. *Biomaterials* 2002;23:4523–4531.

38. Vaandrager JM, van Mullem PJ, de Wijn JR. Porous acrylic cement for the correction of craniofacial deformities and repair of defects, animal experimentation and two years of clinical application. *Biomaterials* 1983;4:128–130.
39. Ramselaar MMA, van Mullein PJ, de Wijn JR. Porous PMMA tooth root replacements: Reactions of the surrounding tissues. *J Dent* 1986;14:202–208.
40. van Mullem PJ, de Wijn JR. Bone and soft connective tissue response to porous acrylic implants: A histokinetic study. *J Cranio-Max Fac Surg* 1988;16:99–109.
41. van Mullem PJ, Vaandrager JM, Nicolai JPA, de Wijn JR. Implantation of porous acrylic cement in soft tissues: An animal and human biopsy histological study. *Biomaterials* 1990;11:299–304.
42. Knabe C, Driessens FCM, Planell JA, Gildenhaar R, Berger G, Reif D, Fitzner R, Radlanski RJ, Gross U. Evaluation of calcium phosphates and experimental calcium phosphate bone cements using osteogenic cultures. *J Biomed Mater Res* 2000;52:498–508.
43. Yuan H, Yang Z, de Bruijn JD, de Groot K, Zhang X. Material-dependent bone induction by calcium phosphate ceramics: A 2.5-year study in dog. *Biomaterials* 2001;22:2617–2623.
44. Kretlow JD, Shi M, Young S, Spicer PP, Demian N, Jansen JA, Wong ME, Kasper FK, Mikos AG. Evaluation of soft tissue coverage over porous polymethylmethacrylate space maintainers within nonhealing alveolar bone defects. *Tissue Eng Part C* 2010;16(6):1427–1438.
45. Shi M, Kretlow JD, Nguyen A, Young S, Baggett LS, Wong ME, Kasper FK, Mikos AG. Antibiotic-releasing porous polymethylmethacrylate constructs for osseous space maintenance and infection control. *Biomaterials* 2010;31:4146–4156.
46. Ishihara K, Arai H, Nakabayashi N, Morita S, Furuya K. Adhesive bone cement containing hydroxyapatite particle as bone compatible filler. *J Biomed Mater Res* 1992;26:937–945.
47. Heikkila JT, Aho AJ, Kangasniemi I, Yli-Urpo A. Polymethylmethacrylate composites: Disturbed bone formation at the surface of bioactive glass and hydroxyapatite. *Biomaterials* 1996;17:1755–1760.
48. Goto K, Tamura J, Shinzato S, Fujibayashi S, Hashimoto M, Kawashita M, Kokubo T, Nakamura T. Bioactive bone cement containing nano-sized titania particles for use as bone substitutes. *Biomaterials* 2005;26:6496–6505.

49. Goto K, Hashimoto M, Takadama H, Tamura J, Fujibayashi S, Kawanabe K, Kokubo T, Nakamura T. Mechanical, setting, and biological properties of bone cements containing micron-sized titania particles. *J Mater Sci Mater Med* 2008;19:1009–1016.
50. Goto K, Shinzato S, Fujibayashi S, Tamura J, Kawanabe K, Hasagawa S, Kowalski R, Nakamura T. The biocompatibility and osteoconductivity of a cement containing β-TCP for use in vertebroplasty. *J Biomed Mater Res* 2006;78(3):629–637.
51. Goto K, Kawanabe K, Kowalski R, Baker D, Nakamura T. Bonding ability evaluation of bone cement on the cortical surface of rabbit's tibia. *J Mater Sci Mater Med* 2010;21:139–146.
52. Fukuda C, Goto K, Imamura M, Nakamura T. Bioactive bone cement with a low content of titania particles without postsilanization: Effect of filler content on osteoconductivity, mechanical properties, and handling characteristics. *J Biomed Mater Res Part B: Appl Biomater* 2010;95(2):407–413.
53. Lye KW, Tideman H, Wolke JCG, Merkx MAW, Chin FKC, Jansen JA. Biocompatibility and bone formation with porous modified PMMA in normal and irradiated mandibular tissue. *Clin Oral Impl Res* 2011;24:100–109.
54. Miyazaki T, Ohtsuki C, Kyomoto M, Tanihara M, Mori A, Kuramoto K. Bioactive PMMA bone cement prepared by modification with methacryloxypropyltrimethoxysilane and calcium chloride. *J Biomed Mater Res* 2003;67(4):1417–1423.
55. Mori A, Ohtsuki C, Miyazaki T, Sugino A, Tanihara M, Kuramoto K, Osaka A. Synthesis of bioactive PMMA bone cement via modification with methacryloxypropyltrimethoxysilane and calcium acetate. *J Mater Sci Mater Med* 2005;16:713–718.
56. Tsukeoka T, Suzuki M, Ohtsuki C, Sugino A, Tsuneizumi Y, Miyagi J, Kuramoto K, Moriya H. Mechanical and histological evaluation of a PMMA-based bone cement modified with γ-methacryloxypropyltrimethoxysilane and calcium acetate. *Biomaterials* 2006;27:3897–3903.
57. Sugino A, Miyazaki T, Kawachi G, Kikuta K, Ohtsuki C. Relationship between apatite-forming ability and mechanical properties of bioactive PMMA-based bone cement modified with calcium salts and alkoxysilane. *J Mater Sci Mater Med* 2008;19:1399–1405.
58. Buchholz HW, Engelbrecht H. Uber die Depotwirkung einiger Antibiotica bei Vermischung mit dem Kunstharz Palacos. *Chirurg* 1970;40:511–515.

59. Andollina A, Bertoni F, Zolezzi C, Trentani F, Trentani P, Borrelli AM, Tigani D. Vancomycin and meropenem in acrylic cement: Elution kinetics of *in vitro* bactericidal action. *Chir Organi Mov* 2008;91: 153–158.
60. Lautenschlager EP, Marshall GW, Marks KE, Schwartz J, Nelson CL. Mechanical strength of acrylic bone cements impregnated with antibiotics. *J Biomed Mater Res* 1976;10:837–845.
61. Lautenschlager EP, Jacobs JJ, Marshall GW, Meyer PR. Mechanical properties of bone cements containing large doses of antibiotic powders. *J Biomed Mater Res* 1976;10:929–938.
62. Buchholz HW, Elson RA, Heinert K. Antibiotic-loaded acrylic cement–current concepts. *Clin Orthop Relat Res* 1984;190:96–108.
63. Wahlig H, Dingeldein E, Buchholz HW, Buchholz M, Bechmann F. Pharmacokinetic study of gentamicin-loaded cement in total hip replacements — comparative effects of varying dosage. *J Bone Joint Sur Br* 1984;66(2):175–179.
64. Klemm K. The use of antibiotic-containing bead chains in the treatment of chronic bone infections. *Clin Microbiol Infect* 2001;7:28–31.
65. Hanssen AD. Prophylactic use of antibiotic bone cement. An emerging standard — in opposition. *J Arthroplasty* 2004;19(4):73–77.
66. Arora MA, Chan EKS, Gupta S, Diwan AD. Polymethylmethacrylate bone cements and additives: A review of the literature. *World J Orthop* 2013;4(2):67–74.
67. Marks KE, Nelson CL, Lautenschlager EP. Antibiotic-impregnated acrylic bone cement. *J Bone Joint Surg Am* 1976;58(3):358–364.
68. Chohfi M, Langlais F, Fourastier J, Minet J, Thomazeau H, Cormier M. Pharmacokinetics, uses, and limitations of vancomycin-loaded bone cement. *Int Orthop* 1998;22:171–177.
69. Bertazzoni ME, Caveiari E, Benini A. Release of antibiotics from polymethylmethacrylate cement. *J Chemotherapy* 2002;14(5):492–500.
70. Yang Y-H, Wang J-C, Pei Y-L. Adding vancomycin to bone cement: Research on its influence on mechanical and fixation strength using rabbit femoral prostheses. *Orthop Surg* 2011;3(4):265–267.
71. Penner MJ, Masri BA, Duncan CP. Elution characteristics of vancomycin and tobramycin combined in acrylic bone–cement. *J Arthroplasty* 1996;11(8):939–944.
72. Masri BA, Duncan CP, Beauchamp CP. Long-term elution of antibiotics from bone–cement. *J Arthroplasty* 1998;13(3):331–338.
73. Della Valle AG, Bostrom M, Brause B, Harney C, Salvati EA. Effective bacterial activity of tobramycin and vancomycin eluted from acrylic bone cement. *Acta Orthop Scandinavica* 2001;72(3):237–240.

74. Morones JR, Elechiguerra JL, Camacho A, Holt K, Kouri JB, Ramirez JT, Yacaman MJ. The bactericidal effect of silver nanoparticles. *Nanotechnology* 2005;16(10):2346–2353.
75. Alt V, Bechert T, Steinrucke P, Wagener M, Seidel P, Dingeldein E, Domann E, Schnettler R. An *in vitro* assessment of the antibacterial properties and cytotoxicity of nanoparticulate silver bone cement. *Biomaterials* 2004;25:4383–4391.
76. Shi Z, Neoh KG, Kang ET, Wang W. Antibacterial and mechanical properties of bone cement impregnated with chitosan nanoparticles. *Biomaterials* 2006;27:2440–2449.
77. Monma H, Kanazawa T. The hydration of α-tricalcium phosphate. *Yogyo-Kyokai-Shi* 1976;84:209–213.
78. Brown WE, Chow LC. A new calcium phosphate, water-setting cement, in Brown PW, editor. *Cement Research Progress*. Westerville, USA: The American Ceramics Society; 1986; pp. 351–379.
79. Lemaitre J, Mirtchi A, Mortier A. Calcium phosphate cements for medical use: State of the art and perspectives of development. *Sil Ind Ceram Sci Technol* 1987;52:141–146.
80. Dorozkin SV. Self-setting calcium orthophosphate formulations: Cements, concretes, pastes and putties. *Int J Mater Chem* 2011;1(1): 1–48.
81. Zhang JT, Tancret F, Bouler JM. Fabrication and mechanical properties of calcium phosphate cements (CPC) for bone substitution. *Mater Sci Eng C* 2011;31:740–747.
82. Tsuru K, Ruslin M, Matsuya S, Ishikawa K. Effects of the method of apatite seed crystals addition on setting reaction of α-tricalcium phosphate based apatite cement. *J Mater Sci Mater Med* 2015;26:244.
83. Chow LC, Hirayama S, Takagi S, Parry E. Diametral tensile strength and compressive strength of a calcium phosphate cement: Effect of applied pressure. *J Biomed Mater Res (Appl Biomater)* 2000;53: 511–517.
84. Phillips RW. *Skinner's Science of Dental Materials*, 9$^{th}$ edition. Philadelphia, PA: W.B. Saunders Company; 1991; pp. 38–39.
85. Gbureck U, Grolms O, Barralet JE, Grover LM, Thull R. Mechanical activation and cement formation of β-tricalcium phosphate. *Biomaterials* 2003;24:4123–4131.
86. Miyamoto Y, Ishikawa K, Fukao H, Sawada M, Nagayama M, Kon M, Asaoka K. *In vivo* setting behavior of fast-setting calcium phosphate cement. *Biomaterials* 1995;16:855–860.

87. Ishikawa K, Takagi S, Chow LC, Ishikawa Y. Properties and mechanisms of fast-setting calcium phosphate cements. *J Mater Sci Mater Med* 1995;6:528–533.
88. Ikenaga M, Hardouin P, Lemaitre J, Andrianjatovo H, Flautre B. Biomechanical characterization of a biodegradable calcium phosphate hydraulic cement: A comparison with porous biphasic calcium phosphate ceramics. *J Biomed Mater Res* 1998;40:139–144.
89. Ishikawa K. Bone substitute fabrication based on dissolution-precipitation reactions. *Materials* 2010;3:1138–1155.
90. Mirtchi J, Lemaitre J, Terao N. Calcium phosphate cements: Study of the β-tricalcium phosphate–monocalcium phosphate system. *Biomaterials* 1989;10:475–480.
91. Bohner M, Lemaitre J. Hydraulic properties of tricalcium phosphate–phosphoric acid–water mixtures, in Duran P, Fernandez JF, editors. *Third Euro-Ceramics*. Castellon de la Plana, Spain: Faenza Editrice Iberica S.L.; 1993; pp. 95–100.
92. Miyamoto Y, Ishikawa K, Takechi M, Toh T, Yuasa T, Nagayama M, Suzuki K. Histological and compositional evaluations of three types of calcium phosphate cements when implanted in subcutaneous tissue immediately after mixing. *J Biomed Mater Res (Appl Biomater)* 1999; 48:36–42.
93. Unezaki Y, Eto T, Inoue H, Minamigawa K, Sugawara F. A study of hardening material consisted of tetracalcium phosphate, dicalcium phosphate dihydrate and citric acid. *J Jpn Prosthodont Soc* 1993;37: 61–66.
94. Unezaki Y, Ryumon H, Inoue H, Okuda H, Onishi H, Minamigawa K. Histocompatibility of a bone graft cement consisting of tetracalcium phosphate dicalcium phosphate dehydrate and citric acid. *Bioceramics* 1996;9:251–253.
95. Doi Y, Takezawa Y, Shibata Y, Wakamatsu N, Kamemizu N, Goto H, Iijima M, Moriwaki Y, Uno Y, Kubo F, Haeuchi W. Self-setting apatite cement: Hydroxyapatite as setting accelerator. *J Dent Mater* 1987;6: 426–431.
96. Liu C, Shen W. Effect of crystal seeding on the hydration of calcium phosphate cement. *J Mater Sci Mater Med* 1997;8:803–807.
97. Yang Q, Troczynski T, Liu D. Influence of apatite seeds on the synthesis of calcium phosphate cement. *Biomaterials* 2002;23:2751–2760.
98. Tsuru K, Ruslin M, Maruta M, Matsuya S, Ishikawa K. Effects of the method of apatite seed crystals addition on setting reaction of

α-tricalcium phosphate based apatite cement. *J Mater Sci Mater Med* 2015; 26:244.

99. Mirtchi AA, Lemaitre J, Munting E. Calcium phosphate cements: Action of setting regulators on the properties of the β-tricalcium phosphate–monocalcium phosphate cements. *Biomaterials* 1989;10: 634–638.
100. Bohner M, Lemaitre J, Ring TA. Effect of sulfate, pyrophosphate, and citrate ions on the physicochemical properties of cements made of β-tricalcium phosphate–phosphoric acid–water mixtures. *J Am Ceram Soc* 1996;79(6):1427–1434.
101. Bohner M, Merkle HP, Van Landuyt P, Trophardy G, Lemaitre J. Effect of several additives and their admixtures on the physico-chemical properties of a calcium phosphate cement. *J Mater Sci: Mater Med* 2000;11:111–116.
102. Ishikawa K, Miyamoto Y, Kon M, Nagayama M, Asaoka K. Non-decay type fast-setting calcium phosphate cement: Composite with sodium alginate. *Biomaterials* 1995;16(7):527–532.
103. Takechi M, Miyamoto Y, Ishikawa K, Yuasa T, Nagayama M, Kon M, Asaoka K. Non-decay type fast-setting calcium phosphate cement using chitosan. *J Mater Sci Mater Med* 1996;7:317–322.
104. Cherng A, Takagi S, Chow LC. Effect of hydroxypropyl methylcellulose and other gelling agents on the handling properties of calcium phosphate cement. *J Biomed Mater Res* 1997;35:273–277.
105. Miyamoto Y, Ishikawa K, Takechi M, Toh T, Yuasa T, Nagayama M, Suzuki K. Basic properties of calcium phosphate cement containing atelocollagen in its liquid or powder phases. *Biomaterials* 1998;19: 707–715.
106. Ishikawa K, Miyamoto Y, Takechi M, Ueyama Y, Suzuki K, Nagayama M, Matsumura T. Effects of neutral sodium hydrogen phosphate on setting reaction and mechanical strength of hydroxyapatite putty. *J Biomed Mater Res* 1999;44(3):322–329.
107. Ishikawa K, Matsuya S, Nakagawa M, Udho K, Suzuki S. Basic properties of apatite cement containing spherical phosphate made with plasma melting method. *J Mater Sci Mater Med* 2004;15:13–17.
108. Thomson RC, Yaszemski MJ, Powers JM, Mikos AG. Hydroxyapatite fibre reinforced poly(a-hydroxy ester) foams for bone regeneration. *Biomaterials* 1998;19:1935–1943.
109. Kasai T, Ishikawa K, Suzuki K, Yatani H. Initial evaluation of ceramic form as a reconstructive material for bone defect. *Dent Mater J* 2000; 19(4):381–388.

110. Pilliar RM, Filiaggi MJ, Wells JD, Grynpas MD, Kandel RA. Porous calcium polyphosphate scaffolds for bone substitute applications — *in vitro* characterization. *Biomaterials* 2001;22:963–972.
111. Karageorgiou V, Kaplan D. Porosity of 3D biomaterial scaffold and osteogenesis. *Biomaterials* 2005;26:5474–5491.
112. Munar ML, Udoh K, Ishikawa K, Matsuya S, Nakagawa M. Effects of sintering temperature over 1300°C on the physical and compositional properties of porous hydroxyapatite foam. *Dent Mater J* 2006; 25(1):51–58.
113. Belyakov AV, Lukin ES, Safronova TV, Safina MN, Pulyaev VI. Porous materials made from calcium phosphate (review). *Glass Ceram* 2008;65(9–10):337–339.
114. Takagi S, Chow LC. Formation of macropores in calcium phosphate cement implant. *J Mater Sci Mater Med* 2001;12:135–139.
115. Xu HHK, Quinn JB, Takagi S, Chow LC, Eichmiller FC. Strong and macroporous calcium phosphate cement: Effect of porosity and fiber reinforcement on mechanical properties. *J Biomed Mater Res* 2001;57:457–466.
116. Tajima S, Kishi Y, Oda M, Maruta M, Matsuya S, Ishikawa K. Fabrication of biporous low-crystalline apatite based on mannitol dissolution from apatite cement. *Dent Mater J* 2006;25:616–620.
117. Xu HHK, Carey LE, Simon CG, Takagi S, Chow LC. Premixed calcium phosphate cements: Synthesis, physical properties, and cell cytotoxicity. *Dent Mater* 2007;23(4):433–441.
118. Xu HHK, Simon Jr, CG. Self-hardening calcium phosphate cement-mesh composite: Reinforcement, macropores, and cell response. *J Biomed Mater Res* 2004;69(2):267–278.
119. Xu HHK, Simon Jr, CG. Self-hardning calcium phosphate composite scaffold for bone tissue engineering. *J Orthop Res* 2004;22:535–543.
120. Xu HHK, Weir MD, Burguera EF, Fraser AM. Injectable and macroporous calcium phosphate cement scaffold. *Biomaterials* 2006;27: 4279–4287.
121. Zhang Y, Xu HHK, Takagi S, Chow LC. *In-situ* hardening hydroxyapatite-based scaffold for bone repair. *J Mater Sci Mater Med* 2006;17: 437–445.
122. Almirall A, Larrecq G, Delgado JA, Planell JA, Ginebra MP. Fabrication of low temperature macroporous hydroxyapatite scaffolds by foaming and hydrolysis of an α-TCP paste. *Biomaterials* 2004;25: 3671–3680.

123. del Valle S, Mino N, Munoz F, Gonzalez A, Planell JA, Ginebra M-P. *In vivo* evaluation of an injectable macroporous calcium phosphate cement. *J Mater Sci Mater Med* 2007;18:353–361.
124. Cama G, Barberis F, Botter R, Cirillo P, Capurro M, Quarto R, Scaglione S, Finocchio E, Mussi V, Valbusa U. Preparation and properties of macroporous brushite bone cements. *Acta Biomater* 2009;5:2161–2168.
125. Shariff KA, Tsuru K, Ishikawa K. Fabrication of interconnected pore forming α-tricalcium phosphate foam granules cement. *J Biomater Appl*, 2016;30:838–845.
126. Udoh K, Munar M, Maruta M, Matsuya S, Ishikawa K. Effects of sintering temperature on physical and compositional properties of α-tricalcium phosphate foam. *Dent Mater J* 2010;29(2):154–159.
127. Wakae H, Takeuchi A, Udoh K, Matsuya S, Munar ML, LeGeros RZ, Nakashima A, Ishikawa K. Fabrication of macroporous carbonate apatite foam by hydrothermal conversion of α-tricalcium phosphate in carbonate solutions. *J Biomed Mater Res A* 2008;87:957–963.
128. Takeuchi A, Munar ML, Wakae H, Maruta M, Matsuya S, Tsuru K, Ishikawa K. Effect of temperature on crystallinity of carbonate apatite foam prepared from α-tricalcium phosphate by hydrothermal treatment. *Biomed Mater Eng* 2009;19:205–211.
129. Schneider G, Blechschmidt K, Linde D, Litschko P, Korbs T, Beleites E. Bone regeneration with glass ceramic implants and calcium phosphate cements in a rabbit cranial defect model. *J Mater Sci Mater Med* 2010;21:2853–2859.
130. Shindo ML, Constantino PD, Friedman CD, Chow LC. Facial skeletal augmentation using hydroxyapatite cement. *Arch Otolaryngol Head Neck Surg* 1993;119:185–190.
131. Constantino PD, Friedman CD, Jones K, Chow LC, Sisson GA. Experimental hydroxyapatite cement cranioplasty. *Plast Reconstr Surg* 1992;90:174–185.
132. Cancian DCJ, Vieira EH, Marcantonio RAC, Garcia Jr, IR. Utilization of autogenous bone, bioactive glasses, and calcium phosphate cement in surgical mandibular bone defects in *Cebus apella* monkeys. *Int J Oral Maxillofac Implants* 2004;19:73–79.
133. Gosain AK, Riordan PA, Song L, Amarante MT, Kalantarian B, Nagy PG, Wilson CR, Toth JM, McIntyre BL. A 1-year study of osteoinduction in hydroxyapatite-derived biomaterials in an adult sheep model: Part II. Bioengineering implants to optimize bone

replacement in reconstruction of cranial defects. *Plast Reconstr Surg* 2004;114:1155–1163.

134. Friedman CD, Costantino PD, Takagi S, Chow LC. BoneSource™ hydroxyapatite cement: A novel biomaterial for craniofacial skeletal tissue engineering and reconstruction. *J Biomed Mater Res* 1998;43: 428–432.
135. Miyamoto Y, Ishikawa K, Takechi M, Toh T, Yoshida Y, Nagayama M, Kon M, Asaoka K. Tissue response to fast-setting calcium phosphate cement in bone. *J Biomed Mater Res* 1997;37:457–464.
136. Ishikawa K. Calcium phosphate cement, in Kokubo T, editor. *Bioceramics and Their Clinical Applications*. New York: CRC Press; 2008; pp. 438–463.
137. Cahyanto A, Tsuru K, Ishikawa K. Carbonate apatite formation during setting reaction of apatite cement. *Adv Bioceram Porous Ceram V: Ceram Eng Sci Proc* 2012;33(6):7–10.
138. LeGelos RZ. Calcium phosphates in oral biology and medicine, in Myers HM, editor. *Monographs in Oral Science*, Vol. 15, Basel: Karger; 1991; p. 110.
139. Leewenburgh S, Layrolle P, Barrère F, de Bruijn J, Schoonman J, van Blitterswijk CA, de Groot K. Osteoclastic resorption of biomimetic calcium phosphate coatings *in vitro*. *J Biomed Mater Res* 2001;56:208–215.
140. Takagi S, Chow LC, Ishikawa K. Formation of hydroxyapatite in new calcium phosphate cements. *Biomaterials* 1998;19:1593–1599.
141. De Leeuw NH, Parker SC. Surface structure and morphology of calcium carbonate polymorphs calcite, aragonite, and vaterite: An atomistic approach. *J Phys Chem B* 1998;102:2914–2922.
142. Plummer LN, Busenberg E. The solubilities of calcite, aragonite and vaterite in $CO_2$–$H_2O$ solutions 0 and 90°C, and an evaluation of the aqueous model for the system $CaCO_3$–$CO_2$–$H_2O$. *Geochem Cosmochim Acta* 1982;46:1011–1040.
143. Combes C, Bareille R, Rey C. Calcium carbonate–calcium phosphate mixed cement compositions for bone reconstruction. *J Biomed Mater Res* 2006;79(2):318–328.
144. Cahyanto A, Michito M, Tsuru K, Matsuya S, Ishikawa K. Fabrication of bone cement that fully transforms to carbonate apatite. *Dent Mater J* 2015;34(3):394–401.
145. Aizawa M, Haruta Y, Okada I. Development of novel cement processing using hydroxyapatite particles modified with inositol phosphate. *Arch BioCeram Res* 2003;3:134–138.

146. Horiguchi Y, Yoshikawa A, Oribe K, Aizawa M. Fabrication of chelate-setting hydroxyapatite cements from four kinds of commercially-available powder with various shape and crystallinity and their mechanical property. *J Ceram Soc Jpn* 2008;116:50–55.
147. Konishi T, Horiguchi Y, Mizumoto M, Honda M, Oribe K, Morisue H, Ishii K, Toyama Y, Matsumoto M, Aizawa M. Novel chelate-setting calcium-phosphate cements fabricated with wet-synthesized hydroxyapatite powder. *J Mater Sci Mater Med* 2013;24:611–621.
148. Konishi T, Mizumoto M, Honda M, Horiguchi Y, Oribe K, Morisue H, Ishii K, Toyama Y, Matsumoto M, Aizawa M. Fabrication of novel biodegradable α-tricalcium phosphate cement set by chelating capability of inositol phosphate and its biocompatibility. *J Nanomater* 2013;2013:article ID 864374.
149. Konishi T, Takahashi S, Zhuang Z, Nagata K, Mizumoto M, Honda M, Takeuchi Y, Matsunari H, Nagashima H, Aizawa M. Biodegradable β-tricalcium phosphate cement with anti-washout property based on chelate-setting mechanism of inositol phosphate. *J Mater Sci Mater Med* 2013;24:1383–1394.
150. Ohsugi H, Habuto Y, Honda M, Aizawa M, Kanzawa N. Evaluation of the anti-bacterial activity of a novel chelate-setting apatite cement containing lactoferrin. *Key Eng Mater* 2013;529–530:187–191.
151. Honda M, Konishi T, Mizumoto M, Aizawa M. *In vitro* biological evaluation of anti-tumor effect of the chelate-setting hydroxyapatite cement. *Key Eng Mater* 2013;529–530:173–177.
152. Inayama T, Konishi H, Aizawa M, Kanzawa N. Studies on the anti-tumor action of chelate-setting apatite cements. *Key Eng Mater* 2013;529–530:178–182.

# Chapter 7

# Strontium-Containing Mesoporous Bioactive Glass for Regeneration of Osteoporotic Bone and Periodontal Tissue

**Chengtie Wu* and Jiang Chang†**

*State Key Laboratory of High Performance Ceramics and Superfine Microstructure, Shanghai Institute of Ceramics, Chinese Academy of Sciences, Shanghai 200050, People's Republic of China*

**chengtiewu@mail.sic.ac.cn*

*†jchang@mail.sic.ac.cn*

## Abstract

Osteoporosis has been one of the most prevalent bone disorders worldwide. Strontium (Sr) has been demonstrated the ability for preventing bone loss and inhibiting the osteoclast functions. Mesoporous bioactive glass (MBG) has shown distinct biological activity for bone regeneration. It is of great interest to combine the therapeutic effect of Sr ions with regeneration function of MBG to develop a new kind of Sr-containing MBG (Sr-MBG) for osteoporosis-related bone and periodontal regeneration applications. In this chapter, we will give an overview on studies related to preparation and properties of Sr-MBG and its stimulation of osteogenic differentiation of bone-forming cells and periodontal ligament cells, and regeneration of osteoporosis-related bone and periodontal regeneration *in vivo*. It is expected that the study about Sr-MBG may offer a new concept that biomaterials, in the future, may be used for therapeutic agents to replace some drugs clinically, avoiding the potential risks and side effect of some functional drugs in human body.

## 7.1 Introduction

Osteoporosis is one of the most prevalent bone disorders worldwide characterised by low bone mineral density, reduction in bone mass, and poor bone strength, leading to skeletal fragility and susceptibility to fractures.[1] The disease progresses with increased age caused by an imbalance between bone-forming osteoblasts and bone-resorbing osteoclasts.[1] Over 200 million people are estimated to be affected worldwide primarily due to postmenopausal oestrogen deficiency.[2–4] Approximately 50% of 65-year-old postmenopausal white or Asian women will experience at least one osteoporosis-related fracture within their lifetime.[4,5] Osteoporosis disease has significant effect not only on bone, but also on the periodontium. More and more studies show that osteoporosis is believed to affect the condition of the periodontal tissues after therapy, given that it may prevent proper healing and ultimately cause severity of the pre-existing periodontal disease.[6–8] Diminished bone density increases the susceptibility towards periodontal breakdown, further complicating regenerative periodontal procedures.[9] For therapy of osteoporosis disease, there are generally two major pharmacological approaches for the treatment of osteoporosis. One is anabolic agents such as parathyroid hormone that stimulate bone formation; the other is anti-resorptive agents including bisphosphonates, calcitonin, raloxifene, and oestrogen for inhibiting bone resorption,[10,11] in which strontium ranelate has received significant attention for the treatment of osteoporosis in the last several years.[4,12,13] Clinical trials have demonstrated that it increases bone mineral density in the lumbar spine, the femoral neck, and in the total hip of osteoporotic patients.[14–17] The significant advantage of strontium ranelate over other therapies is that it simultaneously stimulates bone formation and decreases bone resorption.[4,17–23] Although strontium ranelate, as a pharmacological agent, can be used for treating osteoporosis and preventing disease progression, it remains difficult to regenerate osteoporosis-related injuries post fractures.[4] Sr has been demonstrated to act by preventing bone loss and inhibiting the osteoclast functions.[17] However, one drawback associated with the use of Sr is its bioavailability. Studies have shown that the bioavailability of orally administered strontium is only <20%, and therefore a local release of Sr ions directly into the bone defect site might be preferential to systemic administration.[24] Thus, doping Sr into bone biomaterials for treatment of osteoporotic bone fractures is therefore of research interest.[25,26]

In the last 40 years, bioactive glasses have shown attractive properties for bone tissue regeneration application by virtue of their osteoconductivity and degradability,[27–32] including melt-derived 45S5® bioglass[33–39] and sol–gel-derived 58S.[40–44] The mechanism behind osteoconductivity for bioactive glass is closely associated with the release of $Ca^{2+}$ ions and the deposition of a carbonated hydroxyapatite layer. The apatite layer forms a strong chemical bond between bioactive glasses and host bone.[39] Further studies have also shown that the Ca- and Si-containing ionic products released from the bioactive glasses contribute to their bioactivity, as both Ca and Si are found to stimulate osteoblast proliferation and differentiation.[33–35,37,38,43] Mesoporous bioactive glass (MBG) has attracted significant attention in recent years.[45–48] A significant feature of MBG, compared to non-mesoporous bioactive glass (NBG), is that it possesses significantly improved specific surface area and nanopore volume, which is evidenced by greatly enhanced apatite-mineralisation ability and degradability.[45,49] These materials are based on a $CaO–SiO_2–P_2O_5$ composition and have a highly ordered mesopore channel structure with a pore size ranging from 5 to 20 nm. Compared to conventional NBG, the MBG possesses a more optimal surface area, pore volume, ability to induce *in vitro* apatite mineralisation in simulated body fluids, and excellent cytocompatibility.[43,45,50–52] Recently, we demonstrated that $CaO–P_2O_5–SiO_2$-system MBG has improved *in vitro* and *in vivo* bioactivity, degradation, and drug-delivery properties, compared to NBG.[53–55] As a bioactive material, MBG has shown great potential for bone regeneration applications.[55] Therefore, it is of great interest to combine the therapeutic effect of Sr ions with regeneration function of MBG to develop a new kind of Sr-containing MBG (Sr-MBG) for osteoporosis-related bone and periodontal regeneration applications. The aim of this chapter is to give an overview on studies related to preparation and properties of Sr-MBG and its stimulation of osteogenic differentiation of bone-forming cells and periodontal ligament cells (PDLCs), and regeneration of osteoporosis-related bone and periodontal regeneration.

## 7.2 Preparation, Characterisation, Controlled Sr Ion Release, and Apatite Mineralisation of Sr-MBG Scaffolds

### 7.2.1 Preparation and characterisation of Sr-MBG scaffolds

The main challenge in preparing Sr-containing MBG scaffolds is that the scaffolds should have large pores (approximately several hundreds of micrometres), well-ordered mesopores (approximately several nanometres), and at the same time contain therapeutic $Sr^{2+}$ ions. It is known that Si element is important to form self-assembled mesoporous structure in MBG. If additional elements, such as Ca and Sr, are incorporated in Si-dominated mesoporous materials, it will compromise their mesoporous structure. Therefore, maintaining the well-ordered mesoporous structure of MBG for high bioactivity of scaffolds, but at the same time incorporating therapeutic $Sr^{2+}$ ions into scaffolds, is a challenging issue. To achieve this target, porous Sr-MBG scaffolds were prepared by controlling Sr contents (molar: 2.5%, 5%, and 10%) in MBG using co-templates of non-ionic block polymer P123 (EO20–PO70–EO20) and polyurethane sponges. P123 is used to produce mesoporous structures (mesopore size: several nanometres) and polyurethane sponges are used to create large pores (large pore size: several hundred of micrometres).[55] The prepared Sr-MBG scaffolds have a highly porous structure with a similar large-pore size ranging roughly from 300 to 500 $\mu$m (Figure 7.1).[55] The Sr contents can be effectively controlled, as a sol–gel method is applied to prepare MBG scaffolds in which the added Sr elements will not be lost. As polyurethane sponge

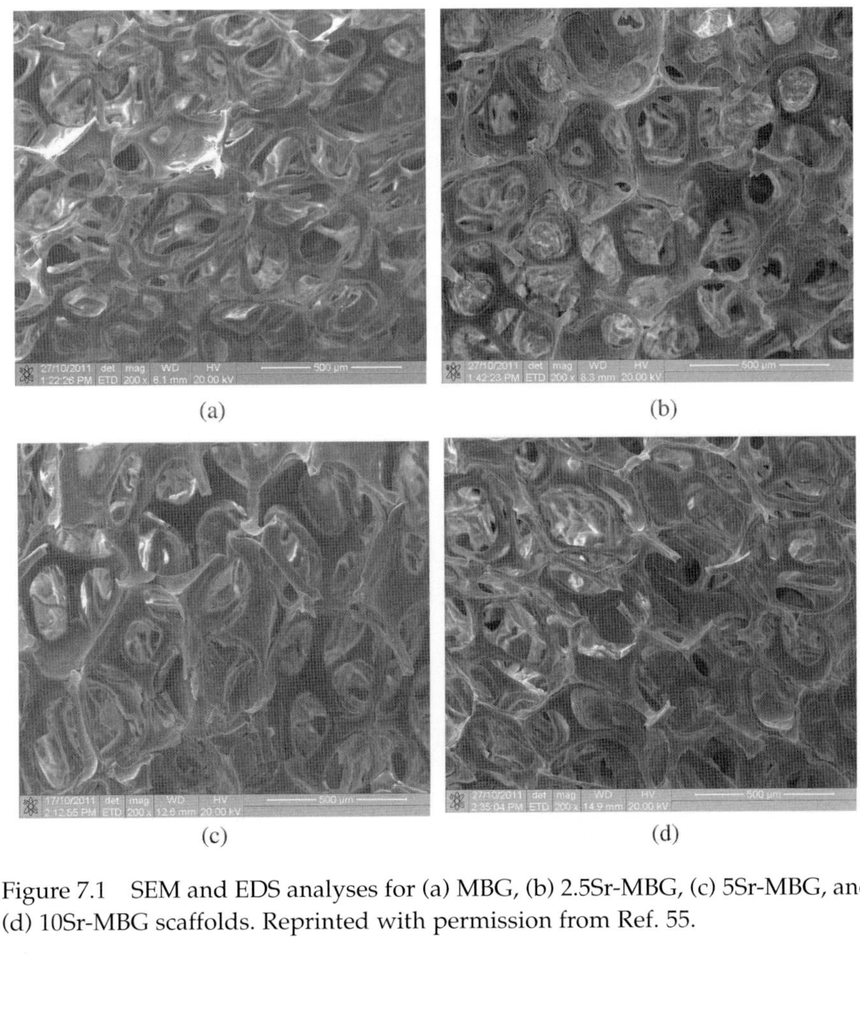


Figure 7.1 SEM and EDS analyses for (a) MBG, (b) 2.5Sr-MBG, (c) 5Sr-MBG, and (d) 10Sr-MBG scaffolds. Reprinted with permission from Ref. 55.

template is used to produce large pores of scaffolds, there is no significant difference for the large-pore porosity among four scaffolds and the porosity of all scaffolds is around 86%, in which the incorporation of Sr will not influence the porosity of the scaffolds.[55] The large pore structure is of importance for cell ingrowths and nutrient transportation. In addition, the prepared Sr-MBG scaffolds maintain well-ordered mesopores. The prepared Sr-MBG scaffold types show a type IV isotherm pattern by $N_2$ adsorption–desorption analysis (Figure 7.2).[55] After incorporating 2.5%, 5%, and 10% of Sr into MBG scaffolds, they still have high specific surface area ranged from 200 to 72 $m^2/g$ and high pore volume ranged from 0.27 to 0.10 $cm^3/g$. Transmission electron microscope (TEM) analysis shows that a well-ordered mesopore can be achieved in the matrix of scaffolds and the mesopore size is around 5 nm (Figure 7.3).[55] Therefore, hierarchically porous Sr-MBG scaffolds have been successfully prepared by using co-templates of polyurethane sponges and P123, in which polyurethane sponges produce large pores for cell

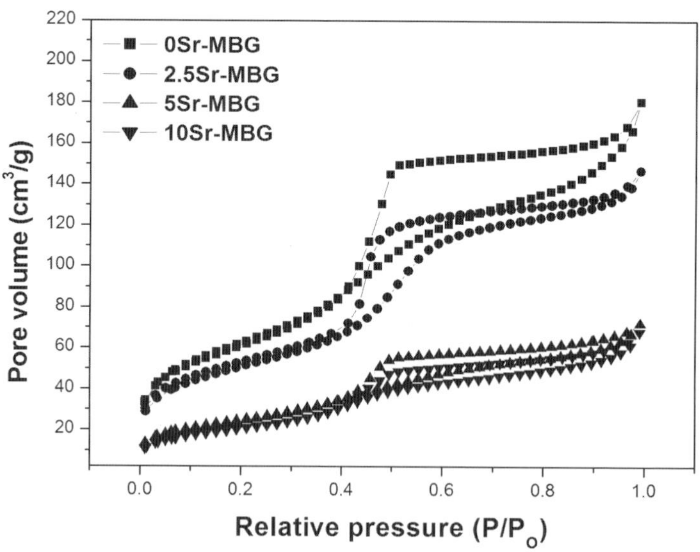


Figure 7.2 Nitrogen adsorption–desorption isotherm analysis for MBG, 2.5Sr-MBG, 5Sr-MBG, and 10Sr-MBG scaffolds. Reprinted with permission from Ref. 55.

(a) (b)

Figure 7.3 TEM analysis for (a) 2.5Sr-MBG and (b) 10Sr-MBG scaffolds. The mesopore size is around 4–5 nm. Reprinted with permission from Ref. 55.

attachments and P123 create mesopores for improving bioactivity and drug delivery.

### 7.2.2 Controlled Sr ion release and apatite mineralisation of Sr-MBG scaffolds

There is a controlled release profile of Sr ions, in which high Sr-containing MBG scaffolds have a quick release of $Sr^{2+}$ ions (Figure 7.4a). Considering the low concentration of Sr in human skeleton (only 3.5 mol.% of Ca content),[56] a controlled $Sr^{2+}$ ion release system is therefore vital for periodontal/bone tissue engineering applications. If the concentrations of $Sr^{2+}$ ions are too high, it will result in potential cytotoxicity associated with health safety problems. Therefore, it is of importance to control the $Sr^{2+}$ ion release from biomaterials. Sr-MBG scaffolds have to satisfy the requirement that the $Sr^{2+}$ ions should have a controlled release by only loading limited amounts of Sr into the scaffolds.[55] It can be seen that the Sr concentrations in the simulated body fluids are less than 24 ppm, a quite low concentration without potential cytotoxicity. Furthermore, Sr-MBG scaffolds have been found to have excellent

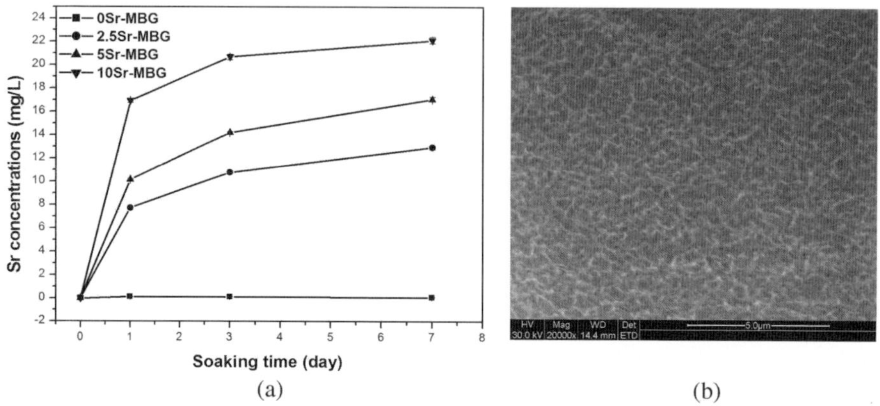


Figure 7.4 Controlled release of Sr ions from Sr-MBG scaffolds (a) and excellent apatite-mineralisation ability of Sr-MBG scaffolds in simulated body fluids (b). Reprinted with permission from Ref. 55.

apatite-mineralisation ability in simulated body fluids, in which nanosized apatite crystals formed on the surface of scaffolds (Figure 7.4b).[55] It is known that apatite mineralisation of bioactive materials plays an important role in maintaining their bioactivity. Apatite mineralisation of biomaterials benefits the integration with host bone.[54,57–60] Furthermore, the formed apatite on the surface of biomaterials possesses the capacity to enhance the osteoblastic activity, including proliferation and differentiation.[55,61–64]

Therefore, Sr has been successfully incorporated into the matrix of MBG scaffolds which still maintain ordered mesoporous structure and large pores as well as apatite-mineralisation ability. Sr-MBG scaffolds have a controlled ability to release Sr ions, which offers a platform in combination of the inherent bioactivity of MBG scaffolds with therapeutic Sr ion release.

## 7.3 Stimulatory Effect of Sr-MBG on the Osteogenic/Cementogenic Differentiation of Both Bone Marrow Stroma Cells (BMSCs) and PDLCs

It is of great significance to investigate the effect of Sr-containing ionic products from Sr-MBG scaffolds on the biological response of ovariectomised (OVX) BMSCs. The results demonstrated that in a moderate range of ion concentration, Sr-containing ionic products from Sr-MBG scaffolds stimulate OVX BMSCs' proliferation. The result also revealed that when the Sr ion concentration was too high, potential cytotoxicity may counteract the stimulatory effect. Therefore, it is of great importance to control the Sr ion release from bioactive scaffolds to satisfy the clinical safety requirements for bone defect repair.[4] Furthermore, it is highly interesting to find that Sr-containing ionic products from Sr-MBG scaffolds was able to significantly increase the gene expression of ALP, COL1, BGLAP, and RUNX2 of OVX BMSCs as assessed by real-time PCR, compared to those from pure MBG scaffolds without Sr (as shown in Figure 7.5).[4] It is known that ALP is an early marker for osteoblastic differentiation.[65–68] COL1 is a major matrix component of bone extracellular matrix and expresses at the beginning of osteoid matrix deposition.[68–70] BGLAP is expressed in the later stage of osteogenic differentiation of bone-forming cells.[67,68] RUNX2 is a master regulator and attractive candidate of target genes for BMP signal in the commitment of osteogenesis.[68,69,71] The results confirmed that the incorporation of Sr into MBG scaffolds could promote osteogenic differentiation of OVX BMSCs.[4]

Sr-MBG scaffolds not only stimulate osteogenic differentiation of BMSCs, but also promote osteogenic/cementogenic differentiation of PDLCs. It is quite interesting to find that the incorporation of parts of Sr into MBG scaffolds significantly stimulated ALP activity and osteogenesis/cementogenesis-related gene expression (ALP, Runx2, OPN, Col I, and CEMP1) of PDLCs (Figure 7.6).[55] PDLCs have generated much attention in periodontal tissue engineering because of their great proliferation ability and multilineage differentiation potential.[72] For this reason, we investigated the interaction of PDLCs with Sr-MBG scaffolds in order to explore whether Sr ions in scaffolds could stimulate the biological response of PDLCs. It is known that CEMP1 is a novel cementum molecule, which is highly expressed by cementoblasts.[73] Being a mineralised tissue, cementum plays a key role in the attachment of connective tissue to the root surface. CEMP1 is found to express at high levels

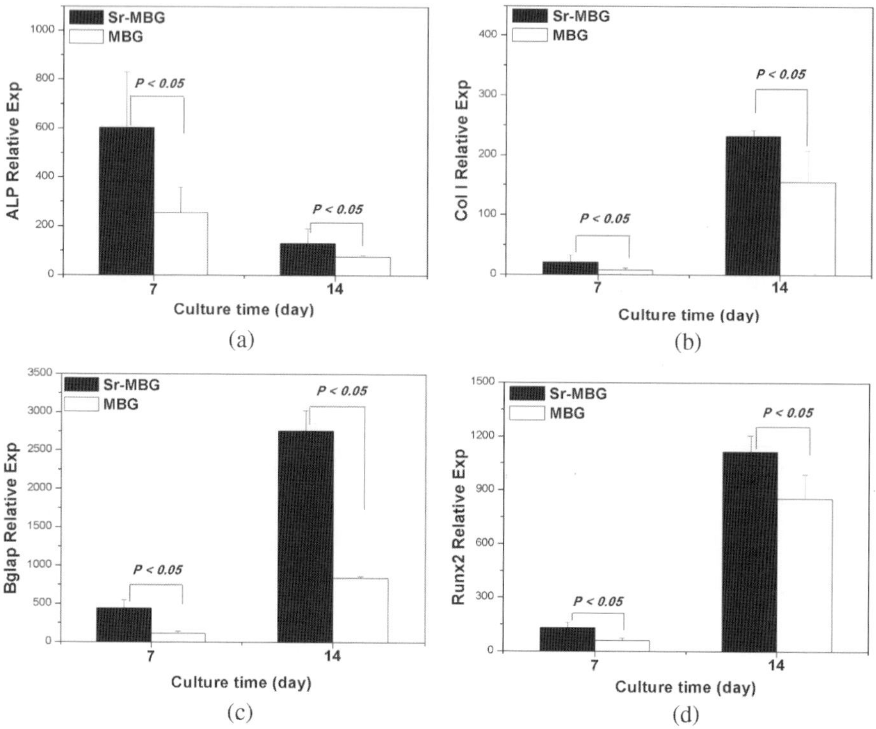


Figure 7.5 Real-time PCR analysis for the bone-related gene expression of OVX rat's BMSCs cultured with Sr-MBG or MBG extracts. The data showed that Sr-MBG presented a significant induction of osteogenic gene, such as ALP (a), COL1 (b), BGALA (c), and RUNX2 (d), which were sequentially expressed in different stages of osteogenesis. Reprinted with permission from Ref. 55.

in cementoblast-like cells and considered as major marker of the cementoblast phenotype.[55,74] Functional cementoblasts also synthesise cementum matrix, which consists of type I collagen as well as other non-collagenous proteins such as osteopontin and osteocalcin. The enhanced CEMP1 expression of PDLCs suggested that the incorporation of Sr into MBG scaffolds stimulated the differentiation towards cementoblast, which is of importance for cementogenesis.[55]

Therefore, the controlled release of Sr ions from Sr-MBG scaffolds has a significant stimulatory effect on both BMSCs and

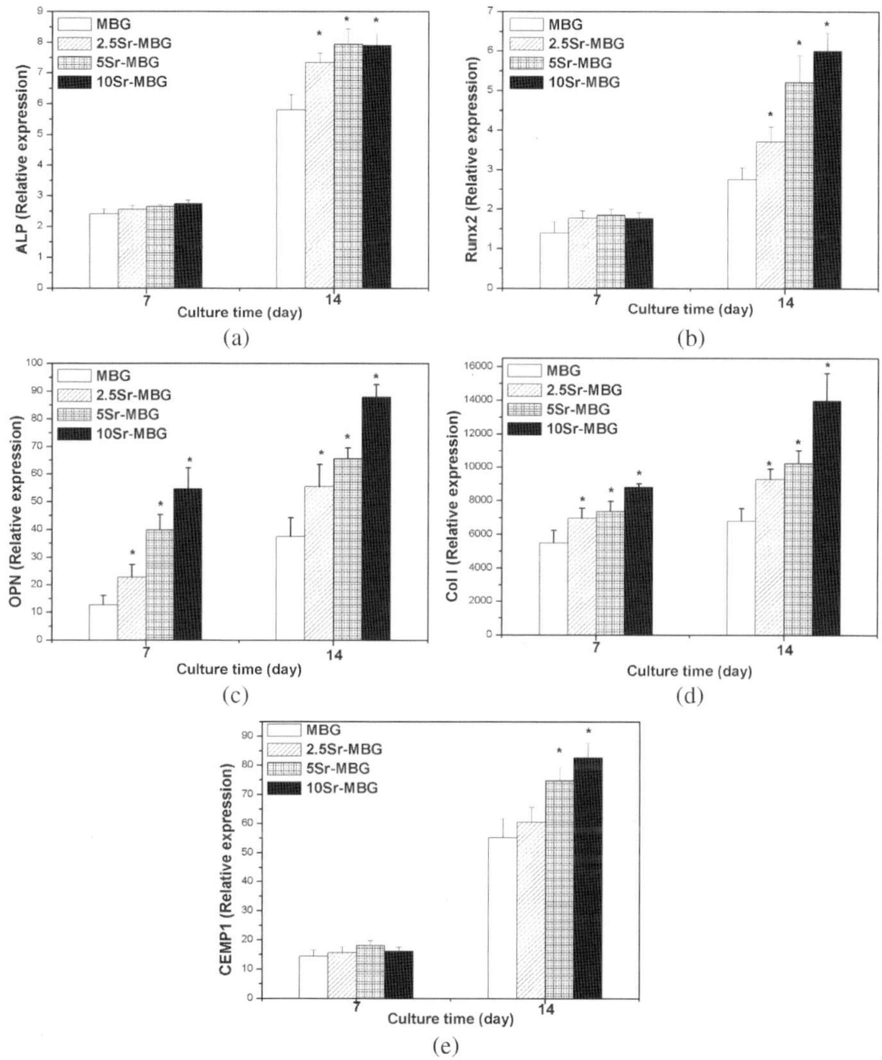


Figure 7.6 The effect of Sr in MBG scaffolds on osteogenic/cementogenic-related gene expression of PDLCs: (a) ALP (a), (b) Runx2, (c) OPN, (d) Col I, and (e) CEMP1 of PDLCs. *Significant difference ($p < 0.05$) for Sr-containing MBG group compared to pure MBG group. Reprinted with permission from Ref. 55.

PDLCs, indicating that Sr and Si ions from MBG may have synergistic effect on the positive effect on cellular proliferation and differentiation. It is known that BMSCs PDLCs are typical tissue cells related to hard tissues, bone, and cementum. The results suggest that the controlled release of Sr ions from MBG scaffolds may have an important effect on the induction of *in vivo* growth of bone and cementum. In addition, one of the important characteristics of third-generation biomaterials is that they can induce specific response to tissue cells, including proliferation and differentiation. Sr-MBG scaffolds have distinctly stimulatory effect on the cellular response (e.g., proliferation, osteogenic/cementogenic gene expression), suggesting that they may be one of third-generation biomaterials. Considered the stimulatory effect of Sr-MBG on both BMSCs and PDLCs, it is reasonable to speculate that Sr-MBG scaffolds could be used for bone and periodontal tissue engineering to treat osteoporosis.

## 7.4 Sr-MBG scaffolds promote the regeneration of osteoporosis-related bone and periodontal defects

In order to verify whether Sr-MBG have ability to regenerate osteoporosis-related bone defects, Sr-MBG scaffolds were implanted in critical sized femur defects in ovariectomy (OVX) rats to mimic the osteoporotic phenotype. The animal study was approved by the Ethics Committee for Animal Research, Wuhan University, China. Mature female Wistar rats (12-week old, mean body weight 200 g) were used for creating osteoporotic phenotype.[4] The rats were subjected to bilateral ovariectomy (OVX) under general anaesthesia by intraperitoneal injection of sodium pentobarbital (10% in weight, 4 ml/kg body weight). Following skin preparation and cutaneous antisepsis, bilateral lumbar lateral incisions which were about 1 cm in length were made, and the enterocoelia was exposed by blunt dissection. The ovaries were pulled out gently and removed post for ligation of the ovarian artery and vein. Then the small incisions were sutured and stratified. Postoperatively, a single intramuscular

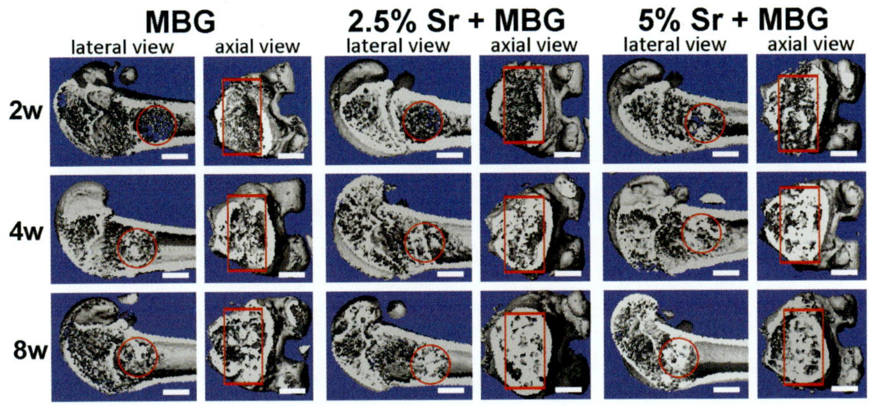


Figure 7.7 3D reconstruction of longitudinal and cross-sectional images by $\mu$CT at 2, 4, and 8 weeks post-operation of the critical femoral defect in OVX rats. The red cycle and rectangle describe the boundary of the defects. Only a few new bones were present in defects at 2 weeks, while abundant new bones regenerated at 4 and 8 weeks with depicted visibly difference among three groups. Scale bar 2 mm. Reprinted with permission from Ref. 4.

dose of penicillin 40,000 IU/mL was administered and there was no sign of inflammation or other notable anomaly. An osteoporosis model was established 2 months later with the diet (18% protein, 0.95% calcium, and 0.65% phosphorus) and water *ad libitum*.[4] After implanted Sr-MBG scaffolds in osteoporotic bone defects, it was found that Sr-MBG scaffolds induced much more mineralisation and higher density than MBG scaffolds alone (Figure 7.7).[4] The morphological and histomorphometry analyses based on the HE and Safranin O staining suggested that Sr-MBG stimulated a more effective osteoconductive and anti-osteoporotic phenotype which increased the speed and quality of bone regeneration and remodelling and a transition of trabecular morphology from a more rod-like appearance to the plate-like appearance with regain of trabecular connectivity. It is worth mentioning that immunohistochemical staining for COL1 further offered the direct evidence of stimulatory effect on OVX BMSCs' osteogenic-related differentiation and osteoid matrix deposition with the help of Sr-MBG scaffolds (Figures 7.8–7.10).[4]

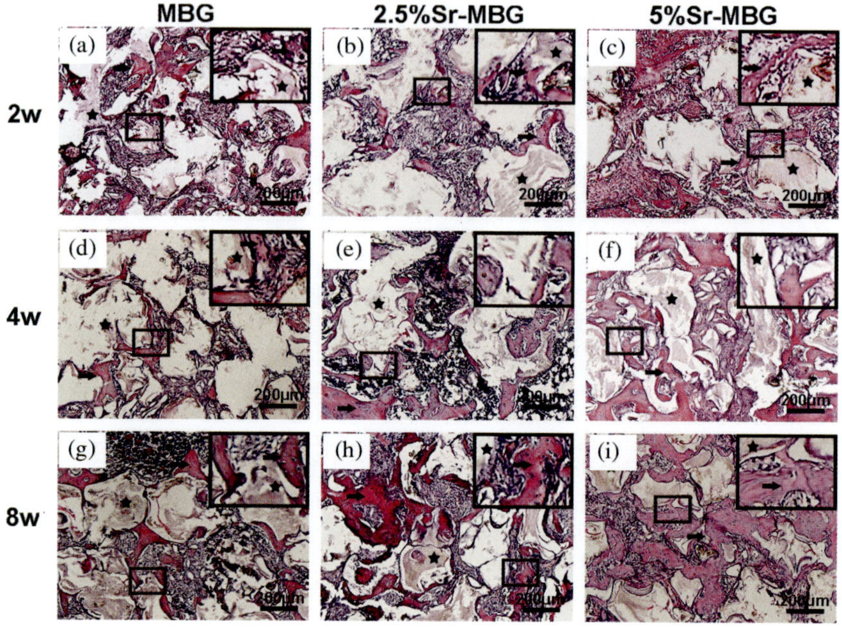


Figure 7.8 Representative sections of HE staining demonstrate the healing defects filled with MBG, 2.5% Sr-MBG, and 5% Sr-MBG in the critical femoral defect of OVX rats. The box in the right top magnifies local details. Arrows: new generated bone, star: filled scaffold, scale bar 200 $\mu$m. Reprinted with permission from Ref. 4.

Further study demonstrated that MBG scaffolds combined with Sr or oestrogen can effectively promote bone regeneration in critical-sized femoral defects in an osteoporotic animal model. Sr elements in MBG scaffolds generate similar ability for forming new bone and inhibiting osteoclast bioactivity for regeneration of osteoporotic bone defects as compared to oestrogen (Figure 7.11).[25] The biocompatibility of the MBG was attributed to both the scaffold components and the bioactive ions loaded together, suggesting an efficient local release profile for Sr or oestrogen can be achieved using MBG scaffolds to further advance bone healing, which is helpful particularly in the treatment of disorders like osteoporosis. The most important fact is that the study offers new evidence that Sr element in biomaterials has comparable functions to regenerate

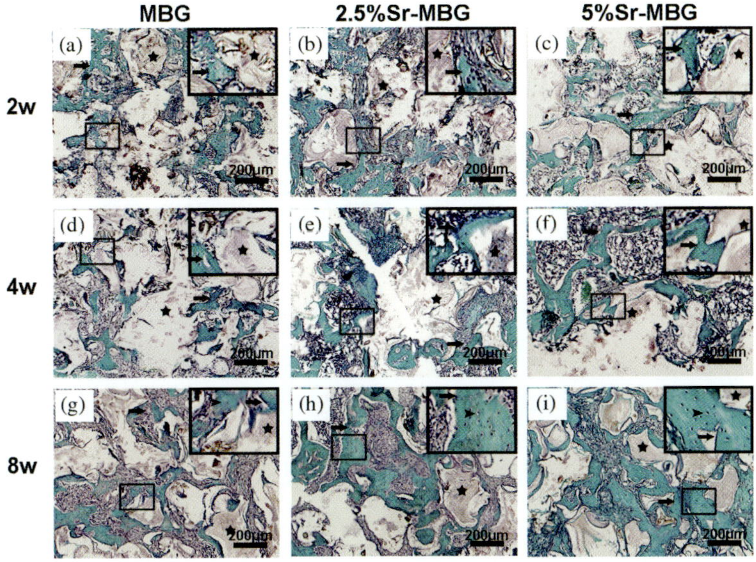


Figure 7.9 Representative sections of Safranin O staining demonstrate the healing defects filled with MBG, 2.5% Sr-MBG, and 5% Sr-MBG in the critical femoral defect of OVX rats. The box in right top magnifies local details. Arrows: new generated bone, arrow head: osteocyte, star: filled scaffold, scale bar 200 $\mu$m. Reprinted with permission from Ref. 4.

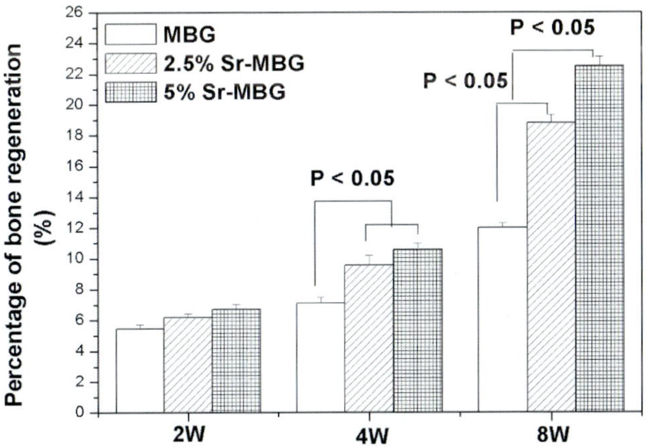


Figure 7.10 Quantitative comparison of the percentage of new bone formation based on the Safranin O staining after implanted with MBG, 2.5% Sr-MBG, and 5% Sr-MBG in the critical femoral defect of OVX rats. Data are presented as mean ± SD and analysed by one-way ANOVA and SNK tests. Sr-MBG induced more new bone formation than MBG scaffolds. Reprinted with permission from Ref. 4.

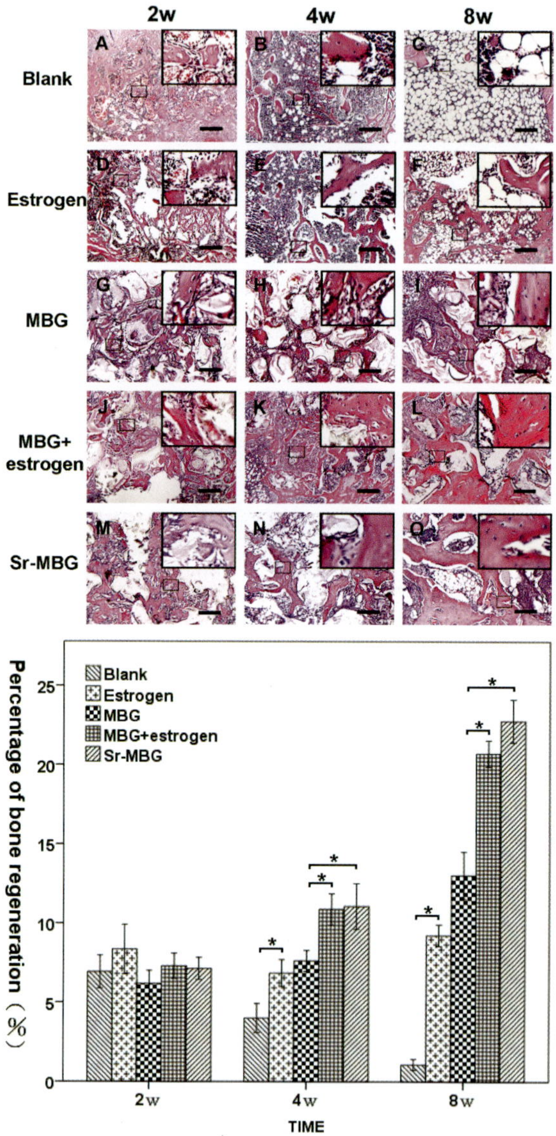


Figure 7.11 The release of Sr ions from MBG scaffolds has comparable ability with oestrogen from MBG scaffolds to regenerate osteoporotic bone defects. Representative sections of H&E staining revealed the osteogenesis among the defect. The box in the top right gave a magnified detail. Bar: 200 $\mu$m. Data were shown as mean $\pm$ SD. $*p < 0.05$. Reprinted with permission from Ref. 25.

osteoporotic bone defects with oestrogen, indicating that Sr-MBG may be used for a therapeutic drug for treating and regenerating osteoporotic bone defects.[25]

Moreover, Sr-MBG scaffolds are capable of significantly enhanced periodontal regeneration in OVX rats post-surgery when compared to control and MBG groups (Figures 7.12 and 7.13).[9] The Sr MBG scaffolds demonstrate that the release of $Sr^{2+}$ ions is capable of improving osteoblast function by increasing new bone formation. Simultaneously, the release of $Sr^{2+}$ ions also significantly decreased the number of multi-nucleated osteoclasts as demonstrated by tartrate resistant acid phosphat (TRAP) staining. Taken together, these results further suggest that the use of Sr-MBG scaffolds may provide greater defect healing in osteoporotic-related periodontal defects.[9]

Finally, it is found that no significant increase in Ca and P ions was observed in either blood or urine samples in animals implanted with Sr-MBG scaffolds versus MBG-alone scaffolds. Although an evident increase in Sr and Si ions was observed, the concentrations were maintained at a very low level which might suggest a slim chance of side effects on overall health. However, the increased concentrations of ions found in blood throughout the body were also observed in urine samples, demonstrating the ability of these animals to secrete abundant ions. Taking the composition of the material itself into consideration, it was reasonable that the concentration of Ca and P increased in urine which collected from scaffolds implanted rats.[4] Previously, it has already been suggested that the calcimimetic effect of Sr may also decrease kidney tubular calcium reabsorption partly leading to the increased calciuria.[75] In addition, it was found that the concentrations of Sr, Si, and Ca ions in urine were higher than those in blood, indicating that the ionic products from Sr-MBG scaffolds can be metabolised and excreted by urine. The findings demonstrate that local delivery of bioactive agents used to treat osteoporosis may also be delivered locally in much lower concentrations to defect areas without having to systematically increase the whole body ion concentration.[4]

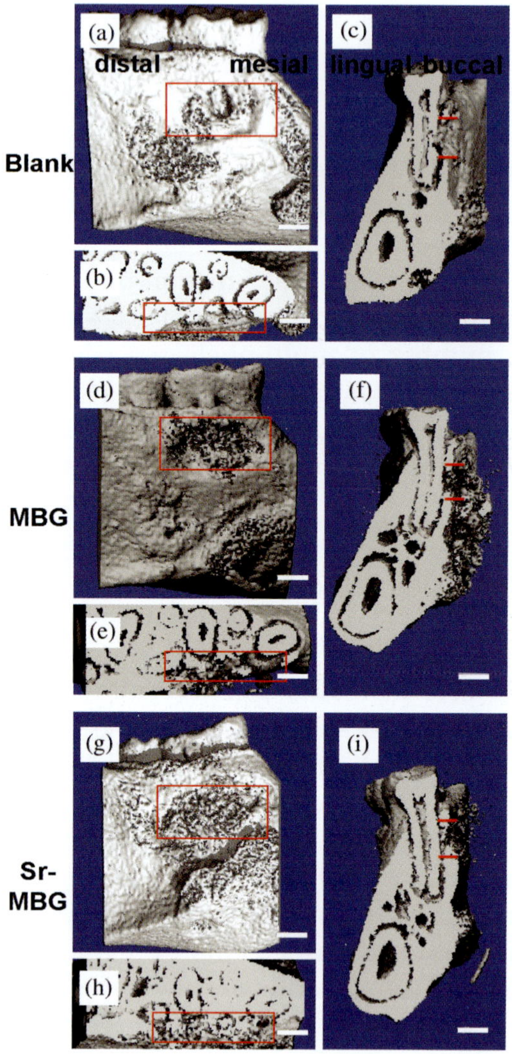


Figure 7.12 The overall state of periodontal regeneration by using Sr-MBG scaffolds was exhibited by 3D reconstruction. The red box represented the original extent of surgical defect in general. A, D, and G: buccal view; B, E, and H: screenshot in horizontal; C, F, and I: screenshot in vertical. Bar: 1 mm. Reprint permission by *PLoS ONE* 2014;9(8):e104527.

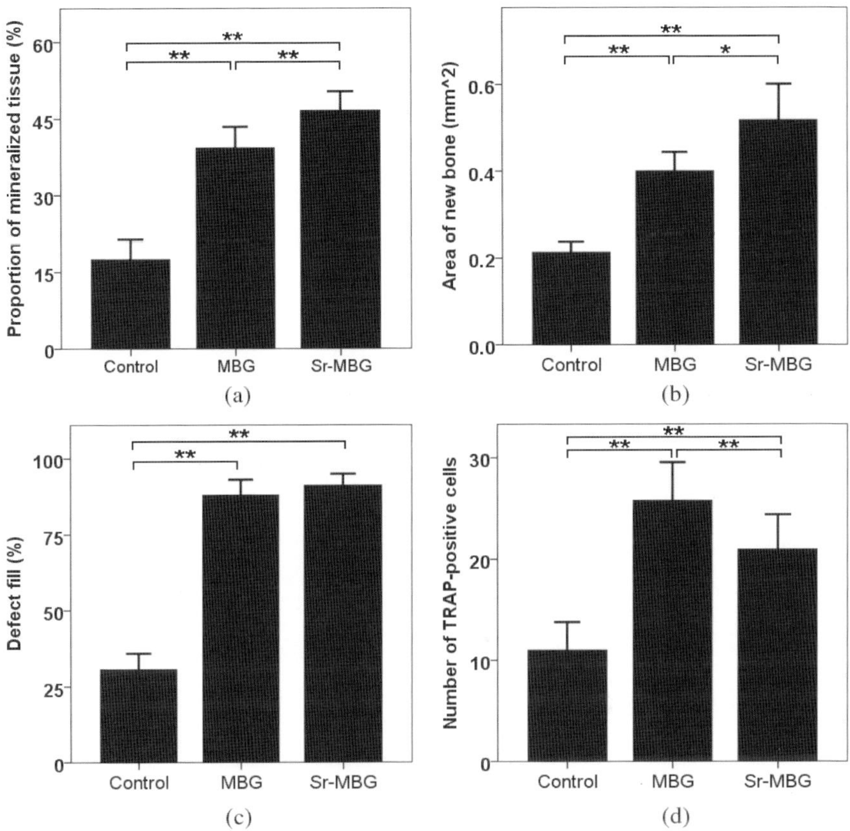


Figure 7.13 Statistical analysis of periodontal regeneration: (a) proportion of mineralised tissue from horizontal cutaway view of 3D construction, (b) area of new bone from H&E stain, (c) defect fill from Masson trichrome stain, and (d) the number of TRAP-positive cells from TRAP stain. Results were given by mean ± SD. $*p < 0.05$, $**p < 0.01$. Reprint permission by *PLoS ONE* 2014;9(8):e104527.

## 7.5 Conclusions and Future Perspectives

In this chapter, we reviewed the state-of-the-art Sr-MBG for delivering therapeutic Sr ions and its stimulatory effect on osteogenic/cementogenic differentiation of both BMSCs and PDLCS as well as regeneration of osteoporotic bone and periodontal tissues. Hierarchical Sr-MBG scaffolds with well-ordered mesopore (~5 nm) and interconnective large pores (~300 $\mu$m) have been successfully prepared by co-templating methods. Sr ions can be released from scaffolds with a controlled profile. The prepared Sr-MBG scaffolds combined the release of therapeutic $Sr^{2+}$ ions and inherent bioactivity of MBG, which leads to a significantly enhanced biological effect on the stimulation of BMSCs and PDLCs and further promotes the regeneration of osteoporotic bone and periodontal tissues. The most important point is that Sr-MBG scaffolds have a comparable ability to regenerate osteoporotic bone as compared to oestrogen, indicating that the bioactive ions (e.g., Sr) in biomaterials can be used as a therapeutic drug for treating and regenerating osteoporotic bone. Further study will be focused on the interactional mechanism of Sr-MBG with stem cells and osteoporotic bone tissues, including large-animal models. However, the study may open a new research direction that biomaterials, in the future, may be used for therapeutic agents to replace some drugs clinically, avoiding the potential risks and side effect of some functional drugs in human body.

## Acknowledgements

This study was funded by the Recruitment Program of Global Young Talent, China (Dr Wu), the National High Technology Research and Development Program of China (863 Program, SS2015AA020302), Natural Science Foundation of China (grants 31370963, 81201202, and 81190132), Key Research Program of Chinese Academy of Sciences (grant KGZD-EW-T06), and Innovative Project of SIC, CAS.

## References

1. Mendonca G, Mendonca DB, Simoes LG, Araujo AL, Leite ER, Duarte WR, *et al*. Nanostructured alumina-coated implant surface: Effect on osteoblast-related gene expression and bone-to-implant contact *in vivo*. *Int J Oral Maxillofac Implants* 2009;24:205–215.
2. Tontonoz P, Pei LM. Fat's loss is bone's gain. *J Clin Invest* 2004; 113:805–806.
3. Rodan GA. Osteopontin overview. *Ann NY Acad Sci* 1995;760:1–5.

4. Zhang YF, Wei LF, Chang J, Miron RJ, Shi B, Yi SQ, *et al.* Strontium-incorporated mesoporous bioactive glass scaffolds stimulating *in vitro* proliferation and differentiation of bone marrow stromal cells and *in vivo* regeneration of osteoporotic bone defects. *J Mater Chem B* 2013;1:5711–5722.
5. Teitelbaum SL. Stem cells and osteoporosis therapy. *Cell Stem Cell* 2010;7:553–554.
6. Passos Jde S, Gomes-Filho IS, Vianna MI, da Cruz SS, Barreto ML, Oliveira TJ, *et al.* Outcome measurements in studies on the association between osteoporosis and periodontal disease. *J Periodontol* 2010;81: 1773–1780.
7. Sultan N, Rao J. Association between periodontal disease and bone mineral density in postmenopausal women: A cross sectional study. *Med Oral Patol Oral Cir Bucal* 2011;16:e440–e447.
8. von Wowern N, Klausen B, Kollerup G. Osteoporosis: A risk factor in periodontal disease. *J Periodontol* 1994;65:1134–1138.
9. Xia L, Lin K, Jiang X, Fang B, Xu Y, Liu J, *et al.* Effect of nano-structured bioceramic surface on osteogenic differentiation of adipose derived stem cells. *Biomaterials* 2014;35:8514–8527.
10. da Silva JM, Andrade Junior CV, Zaia AA, Pessoa OF. Microscopic cleanliness evaluation of the apical root canal after using calcium hydroxide mixed with chlorhexidine, propylene glycol, or antibiotic paste. *Oral Surg Oral Med Oral Pathol Oral Radiol Endod* 2011;111: 260–264.
11. Boonen S, Ferrari S, Miller PD, Eriksen EF, Sambrook PN, Compston J, *et al.* Postmenopausal osteoporosis treatment with antiresorptives: Effects of discontinuation or long-term continuation on bone turnover and fracture risk — A perspective. *J Bone Mineral Res* 2012;27:963–974.
12. Roschger P, Manjubala I, Zoeger N, Meirer F, Simon R, Li C, *et al.* Bone material quality in transiliac bone biopsies of postmenopausal osteoporotic women after 3 years of strontium ranelate treatment. *J Bone Mineral Res* 2010;25:891–900.
13. Li C, Paris O, Siegel S, Roschger P, Paschalis EP, Klaushofer K, *et al.* Strontium is incorporated into mineral crystals only in newly formed bone during strontium ranelate treatment. *J Bone Mineral Res* 2010;25: 968–75.
14. Hamadouche M, Meunier A, Greenspan DC, Blanchat C, Zhong JP, La Torre GP, *et al.* Long-term *in vivo* bioactivity and degradability of bulk sol–gel bioactive glasses. *J Biomed Mater Res* 2001;54:560–566.

15. Reginster JY, Seeman E, De Vernejoul MC, Adami S, Compston J, Phenekos C, *et al.* Strontium ranelate reduces the risk of nonvertebral fractures in postmenopausal women with osteoporosis: Treatment of Peripheral Osteoporosis (TROPOS) study. *J Clin Endocrinol Metab* 2005;90:2816–2822.
16. Seeman E, Devogelaer JP, Lorenc R, Spector T, Brixen K, Balogh A, *et al.* Strontium ranelate reduces the risk of vertebral fractures in patients with osteopenia. *J Bone Mineral Res* 2008;23:433–438.
17. Marie PJ, Ammann P, Boivin G, Rey C. Mechanisms of action and therapeutic potential of strontium in bone. *Calcif Tissue Int* 2001;69:121–129.
18. Verberckmoes SC, De Broe ME, D'Haese PC. Dose-dependent effects of strontium on osteoblast function and mineralization. *Kidney Int* 2003;64:534–543.
19. Qiu K, Zhao XJ, Wan CX, Zhao CS, Chen YW. Effect of strontium ions on the growth of ROS17/2.8 cells on porous calcium polyphosphate scaffolds. *Biomaterials* 2006;27:1277–1286.
20. Peng S, Liu XS, Zhou G, Li Z, Luk KD, Guo XE, *et al.* Osteoprotegerin deficiency attenuates strontium-mediated inhibition of osteoclastogenesis and bone resorption. *J Bone Miner Res* 2011;26:1272–1282.
21. Choudhary S, Halbout P, Alander C, Raisz L, Pilbeam C. Strontium ranelate promotes osteoblastic differentiation and mineralization of murine bone marrow stromal cells: Involvement of prostaglandins. *J Bone Mineral Res* 2007;22:1002–1010.
22. Blake GM, Compston JE, Fogelman I. Could strontium ranelate have a synergistic role in the treatment of osteoporosis? *J Bone Mineral Res* 2009;24:1354–1357.
23. Miron RJ, Zhang YF. Osteoinduction: A review of old concepts with new standards. *J Dent Res* 2012;91:736–744.
24. Kendler DL. Strontium ranelate — Data on vertebral and nonvertebral fracture efficacy and safety: Mechanism of action. *Curr Osteoporos Rep* 2006;4:34–39.
25. Wei L, Ke J, Prasadam I, Miron RJ, Lin S, Xiao Y, *et al.* A comparative study of Sr-incorporated mesoporous bioactive glass scaffolds for regeneration of osteopenic bone defects. *Osteoporos Int* 2014; 25:2089–2096.
26. Wu C, Ramaswamy Y, Kwik D, Zreiqat H. The effect of strontium incorporation into $CaSiO_3$ ceramics on their physical and biological properties. *Biomaterials* 2007;28:3171–3181.

27. Misra SK, Ansari T, Mohn D, Valappil SP, Brunner TJ, Stark WJ, *et al.* Effect of nanoparticulate bioactive glass particles on bioactivity and cytocompatibility of poly(3-hydroxybutyrate) composites. *J R Soc Interf* 2010;7:453–465.
28. Hench LL, Thompson I. Twenty-first century challenges for biomaterials. *J R Soc Interf* 2010;7(Suppl 4):S379–S391.
29. Chen QZ, Thompson ID, Boccaccini AR. 45S5 Bioglass-derived glass-ceramic scaffolds for bone tissue engineering. *Biomaterials* 2006;27: 2414–2425.
30. Jones JR, Tsigkou O, Coates EE, Stevens MM, Polak JM, Hench LL. Extracellular matrix formation and mineralization on a phosphate-free porous bioactive glass scaffold using primary human osteoblast (HOB) cells. *Biomaterials* 2007;28:1653–1663.
31. Jones JR, Ehrenfried LM, Hench LL. Optimising bioactive glass scaffolds for bone tissue engineering. *Biomaterials* 2006;27:964–973.
32. Wu ZY, Hill RG, Yue S, Nightingale D, Lee PD, Jones JR. Melt-derived bioactive glass scaffolds produced by a gel-cast foaming technique. *Acta Biomater* 2011;7:1807–1816.
33. Gough JE, Jones JR, Hench LL. Nodule formation and mineralisation of human primary osteoblasts cultured on a porous bioactive glass scaffold. *Biomaterials* 2004;25:2039–2046.
34. Gough JE, Notingher I, Hench LL. Osteoblast attachment and mineralized nodule formation on rough and smooth 45S5 bioactive glass monoliths. *J Biomed Mater Res A* 2004;68:640–650.
35. Valerio P, Pereira MM, Goes AM, Leite MF. The effect of ionic products from bioactive glass dissolution on osteoblast proliferation and collagen production. *Biomaterials* 2004;25:2941–2948.
36. Xynos ID, Edgar AJ, Buttery LD, Hench LL, Polak JM. Gene-expression profiling of human osteoblasts following treatment with the ionic products of Bioglass 45S5 dissolution. *J Biomed Mater Res* 2001;55:151–157.
37. Xynos ID, Edgar AJ, Buttery LD, Hench LL, Polak JM. Ionic products of bioactive glass dissolution increase proliferation of human osteoblasts and induce insulin-like growth factor II mRNA expression and protein synthesis. *Biochem Biophys Res Commun* 2000; 276:461–465.
38. Hoppe A, Guldal NS, Boccaccini AR. A review of the biological response to ionic dissolution products from bioactive glasses and glass-ceramics. *Biomaterials* 2011;32:2757–2774.

39. Hench LL. Bioceramics: From concept to clinic. *J Am Ceram Soc* 1991;74:1487–1510.
40. Li R, Clark AE, Hench LL. An investigation of bioactive glass powders by sol–gel processing. *J Appl Biomater* 1991;2:231–239.
41. Arcos D, Lopez-Noriega A, Ruiz-Hernandez E, Terasaki O, Vallet-Regi M. Ordered mesoporous microspheres for bone grafting and drug delivery. *Chem Mater* 2009;21:1000–1009.
42. Wu C, Chang J, Xiao Y. Mesoporous bioactive glasses as drug delivery and bone tissue engineering platforms. *Therapeut Deliv* 2011; 2:1189–1198.
43. Wu C, Chang J. Mesoporous bioactive glasses: Structure characteristics, drug/growth factor delivery and bone regeneration application. *Interf Focus* 2012;2:292–306.
44. Wu C, Chang J. Multifunctional mesoporous bioactive glasses for effective delivery of therapeutic ions and drug/growth factors. *J Control Rel* 2014;193:282–295.
45. Yan X, Huang X, Yu C, Deng H, Wang Y, Zhang Z, *et al*. The *in-vitro* bioactivity of mesoporous bioactive glasses. *Biomaterials* 2006;27: 3396–3403.
46. Yan X, Yu C, Zhou X, Tang J, Zhao D. Highly ordered mesoporous bioactive glasses with superior *in vitro* bone-forming bioactivities. *Angew Chem Int Ed Engl* 2004;43:5980–5984.
47. Vallet-Regi MA, Ruiz-Gonzalez L, Izquierdo-Barba I, Gonzalez-Calbet JM. Revisiting silica based ordered mesoporous materials: Medical applications. *J Mater Chem* 2006;16:26–31.
48. Vallet-Regi M, Balas F, Arcos D. Mesoporous materials for drug delivery. *Angew Chem Int Ed Engl* 2007;46:7548–7558.
49. Li X, Shi J, Dong X, Zhang L, Zeng H. A mesoporous bioactive glass/ polycaprolactone composite scaffold and its bioactivity behavior. *J Biomed Mater Res A* 2008;84:84–91.
50. Leonova E, Izquierdo-Barba I, Arcos D, Lopez-Noriega A, Hedin N, Vallet-Regi M, *et al*. Multinuclear solid-state NMR studies of ordered mesoporous bioactive glasses. *J Phys Chem C* 2008;112:5552–5562.
51. Garcia A, Cicuendez M, Izquierdo-Barba I, Arcos D, Vallet-Regi M. Essential role of calcium phosphate heterogeneities in 2D-hexagonal and 3D-cubic $SiO_2$–CaO–$P_2O_5$ mesoporous bioactive glasses. *Chem Mater* 2009;21:5474–5484.
52. Alcaide M, Portoles P, Lopez-Noriega A, Arcos D, Vallet-Regi M, Portoles MT. Interaction of an ordered mesoporous bioactive glass

with osteoblasts, fibroblasts and lymphocytes, demonstrating its biocompatibility as a potential bone graft material. *Acta Biomater* 2010;6:892–899.
53. Wu C, Ramaswamy Y, Zhu Y, Zheng R, Appleyard R, Howard A, *et al.* The effect of mesoporous bioactive glass on the physiochemical, biological and drug-release properties of poly(DL-lactide-co-glycolide) films. *Biomaterials* 2009;30:2199–2208.
54. Wu C, Zhang Y, Zhou Y, Fan W, Xiao Y. A comparative study of mesoporous-glass/silk and non-mesoporous-glass/silk scaffolds: Physicochemistry and *in vivo* osteogenesis. *Acta Biomater* 2011;7:2229–2236.
55. Wu C, Zhou Y, Lin C, Chang J, Xiao Y. Strontium-containing mesoporous bioactive glass scaffolds with improved osteogenic/cementogenic differentiation of periodontal ligament cells for periodontal tissue engineering. *Acta Biomater* 2012;8:3805–3815.
56. Fourman P, Royer P, Levell M, Morgan D. *Calcium Metabolism and the Bone*, 2$^{nd}$ edition. Oxford: Blackwell; 1968.
57. Hench LL. Biomaterials: A forecast for the future. *Biomaterials* 1998; 19:1419–1423.
58. Xue W, Liu X, Zheng X, Ding C. *In vivo* evaluation of plasma-sprayed wollastonite coating. *Biomaterials* 2005;26:3455–3460.
59. Xu S, Lin K, Wang Z, Chang J, Wang L, Lu J, *et al.* Reconstruction of calvarial defect of rabbits using porous calcium silicate bioactive ceramics. *Biomaterials* 2008;29:2588–2596.
60. Wu C, Zhang Y, Fan W, Ke X, Hu X, Zhou Y, *et al.* $CaSiO_3$ microstructure modulating the *in vitro* and *in vivo* bioactivity of poly(lactide-co-glycolide) microspheres. *J Biomed Mater Res A* 2011;98:122–131.
61. Wu C, Chang J, Zhai W, Ni S. A novel bioactive porous bredigite ($Ca_7MgSi_4O_{16}$) scaffold with biomimetic apatite layer for bone tissue engineering. *J Mater Sci Mater Med* 2007;18:857–864.
62. El-Ghannam A, Ducheyne P, Shapiro IM. Formation of surface reaction products on bioactive glass and their effects on the expression of the osteoblastic phenotype and the deposition of mineralized extracellular matrix. *Biomaterials* 1997;18:295–303.
63. Loty C, Sautier JM, Boulekbache H, Kokubo T, Kim HM, Forest N. *In vitro* bone formation on a bone-like apatite layer prepared by a biomimetic process on a bioactive glass-ceramic. *J Biomed Mater Res* 2000;49: 423–434.
64. Olmo N, Martin AI, Salinas AJ, Turnay J, Vallet-Regi M, Lizarbe MA. Bioactive sol–gel glasses with and without a hydroxycarbonate

apatite layer as substrates for osteoblast cell adhesion and proliferation. *Biomaterials* 2003;24:3383–3393.

65. Gharibi B, Abraham AA, Ham J, Evans BA. Adenosine receptor subtype expression and activation influence the differentiation of mesenchymal stem cells to osteoblasts and adipocytes. *J Bone Mineral Res* 2011;26:2112–2124.
66. Otto F, Thornell AP, Crompton T, Denzel A, Gilmour KC, Rosewell IR, *et al.* Cbfa1, a candidate gene for cleidocranial dysplasia syndrome, is essential for osteoblast differentiation and bone development. *Cell* 1997;89:765–771.
67. Kim K, Jeong CG, Hollister SJ. Non-invasive monitoring of tissue scaffold degradation using ultrasound elasticity imaging. *Acta Biomater* 2008;4:783–790.
68. Pratap J, Lian JB, Javed A, Barnes GL, van Wijnen AJ, Stein JL, *et al.* Regulatory roles of Runx2 in metastatic tumor and cancer cell interactions with bone. *Cancer Metastasis Rev* 2006;25:589–600.
69. Assoian RK, Komoriya A, Meyers CA, Miller DM, Sporn MB. Transforming growth factor-beta in human platelets. Identification of a major storage site, purification, and characterization. *J Biol Chem* 1983;258:7155–7160.
70. Hidalgo-Bastida LA, Cartmell SH. Mesenchymal stem cells, osteoblasts and extracellular matrix proteins: Enhancing cell adhesion and differentiation for bone tissue engineering. *Tissue Eng Part B Rev* 2010;16:405–412.
71. Ivey KN, Srivastava D. MicroRNAs as regulators of differentiation and cell fate decisions. *Cell Stem Cell* 2010;7:36–41.
72. Nanci A, Bosshardt DD. Structure of periodontal tissues in health and disease. *Periodontology 2000* 2006;40:11–28.
73. Alvarez-Perez MA, Narayanan S, Zeichner-David M, Rodriguez Carmona B, Arzate H. Molecular cloning, expression and immunolocalization of a novel human cementum-derived protein (CP-23). *Bone* 2006;38:409–419.
74. Paula-Silva FW, Ghosh A, Arzate H, Kapila S, da Silva LA, Kapila YL. Calcium hydroxide promotes cementogenesis and induces cementoblastic differentiation of mesenchymal periodontal ligament cells in a CEMP1- and ERK-dependent manner. *Calcif Tissue Int* 2010;87:144–157.
75. Oste L, Bervoets AR, Behets GJ, Dams G, Marijnissen RL, Geryl H, *et al.* Time-evolution and reversibility of strontium-induced osteomalacia in chronic renal failure rats. *Kidney Int* 2005;67:920–930.

# Chapter 8

# Bioactive Glass-Ceramics

**Taishi Yokoi[*,¶], Toshiki Miyazaki[†,‖], Masakazu Kawashita[‡,**] and Chikara Ohtsuki[§,††]**

[*] *Materials Research and Development Laboratory, Japan Fine Ceramics Center, Japan*

[†] *Graduate School of Life Science and Systems Engineering, Kyushu Institute of Technology, Japan*

[‡] *Department of Biomedical Engineering, Graduate School of Biomedical Engineering, Tohoku University, Japan*

[§] *Department of Crystalline Materials Science, Graduate School of Engineering, Nagoya University, Japan*

[¶] *t_yokoi@jfcc.or.jp, yokoi@mail.kankyo.tohoku.ac.jp;*
[‖] *tmiya@life.kyutech.ac.jp;*

[**] *m-kawa@ecei.tohoku.ac.jp;* [††] *ohtsuki@apchem.nagoya-u.ac.jp*

**Abstract**

Glass-ceramics have been studied as materials for the reconstructions of hard tissues because they display high compatibility with bones (i.e., bioactivity). In this chapter, the following topics are discussed to illustrate the application of bioactive glass-ceramics in biomedical applications: bone-bonding mechanism of glass-ceramics, biological and mechanical properties of clinically applied bioactive glass-ceramics, recent progress in bioactive glass-ceramics, and glass-ceramics applied in areas other than hard-tissue regeneration.

## 8.1 Introduction

The crystallisation of glass enables the development of new glass materials with various functions. Conventional glass is prepared by a melt-quenching technique that involves quenching of a supercooled liquid with a random atomic arrangement. Glass is thermodynamically metastable when compared to a crystalline material. Hence, when glass is heated to temperatures higher than its transition point, the viscosity of the supercooled liquid decreases. Re-arrangement of atoms occurs in the heated supercooled liquid. Subsequently, nucleation and crystal growth occur to form a thermodynamically stable crystal phase. This phenomenon is used towards the fabrication of glass–crystal composite materials. Glass–crystal composites are generally termed as 'glass-ceramics'. The crystal phase and crystal size of the crystalline component can be controlled by adjusting the heat-treatment conditions and glass composition. Thus, glass can be converted into glass-ceramics comprising diverse crystalline phases with controlled crystal size and component by heat treatment. The resulting glass-ceramic material can feature superior properties to the parent glass. For instance, improved mechanical strength is a representative effect resulting from the crystallisation of glass.

Glass-ceramics have been prospective materials as bone substitutes, the so-called artificial bones, for biomedical applications in orthopaedic surgery. The history of modern ceramic biomaterials displaying high biological affinity began with the discovery of bioactive materials by Hench *et al.*[1] in the early 1970s. The term "bioactivity" refers to the phenomenon by which a biomaterial exhibits specific biological activity. Briefly, the term relates to the bone-bonding property of a material in the field of ceramic biomaterials particularly applied in regeneration of hard tissues. Generally, artificial materials are encapsulated within fibrous tissues after implantation in bony defects. Hench *et al.* reported that glass with certain composition ranges within a $Na_2O$–CaO–$SiO_2$–$P_2O_5$ system showed direct bonding to bone after implantation in bony defects. Accordingly, the authors exploited such phenomenon to develop bone-repairing materials. The discovery of bioactive glass led to studies on bioactive materials with high biological affinity for living bone tissues. The bioactive materials include glass materials, ceramics, inorganic–organic composites, and coatings on substrates.

Typically, bioactive glass-ceramics are derived from bioactive glass. They have superior properties than the original bioactive glass. Hence, the development of bioactive glass-ceramics has been an important part in the history of ceramic biomaterials and is expected to remain an important area of study in the future. In this chapter, we review different bioactive glass-ceramics and related materials developed to date and their applications.

## 8.2 Bonding Mechanism of Glass Materials to Bones

Bone-bonding property relates to the osteoconduction phenomenon, whereby bone grows and remodels over a surface of an implanted material. Previous studies on bioactive glass and glass-ceramic materials revealed that bioactive glass materials induce the formation of a Ca,P-rich layer on their surface after implantation in bony defects.[2] This Ca,P-rich layer constitutes of hydroxyapatite (stoichiometric composition: $Ca_{10}(PO_4)_6(OH)_2$).[3,4] The bioactive glass materials bond to bone through the hydroxyapatite layer. Figure 8.1[4] exemplifies the interface between bone and an implanted type of bioactive glass-ceramic that contains apatite, namely oxyfluorapatite ($Ca_{10}(PO_4)_6(O,F_2)$), and $\beta$-wollastonite ($\beta$-$CaSiO_3$). A layer of hydroxyapatite can be observed at the interface between bone and glass-ceramic. Though the glass-ceramic contains oxyfluorapatite and $\beta$-wollastonite, the formation of a hydroxyapatite layer is not observed on the surface of the glass-ceramic without bone-bonding capabilities. Thus, the bone-bonding property of glasses and glass-ceramics is governed by the formation of a hydroxyapatite layer after implantation in the bone

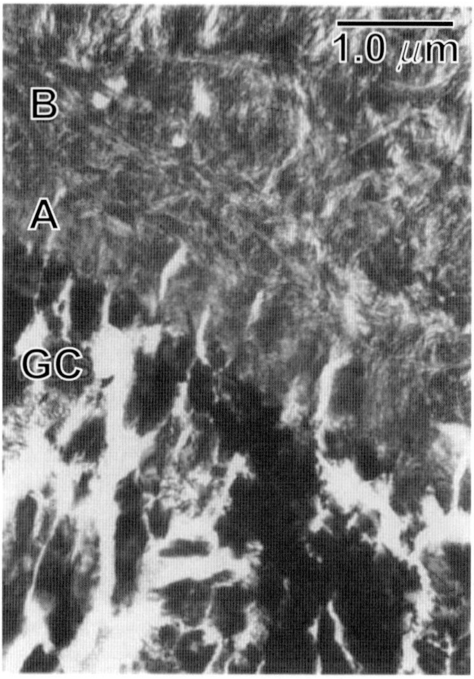


Figure 8.1 Transmission electron microscope (TEM) image of the interface between bone (B) and glass-ceramic containing apatite (GC). GC bonds to bone through intermediate apatite layer (A). Reproduced with permission from Ref. 4. Copyright 1992, Wiley-VCH Verlag GmbH & Co. KGaA.

tissues. The bonding process of bioactive glasses and glass-ceramics to bone is schematically shown in Figure 8.2. The hydroxyapatite-forming ability under *in vivo* conditions is an essential requirement for glasses and glass-ceramics to exhibit bioactivity.

Hydroxyapatite formation on the surface of bioactive glasses and glass-ceramics occurs after soaking in an aqueous solution that has a similar inorganic composition to human blood plasma. The solution prepared by Kokubo *et al.*[5] can be used as a test solution to investigate hydroxyapatite formation on target materials and is commonly referred to as simulated body fluid (SBF). The composition of the inorganic ions constituting SBF is given in Table 8.1. SBF does not contain any biological components. Hence, its preparation

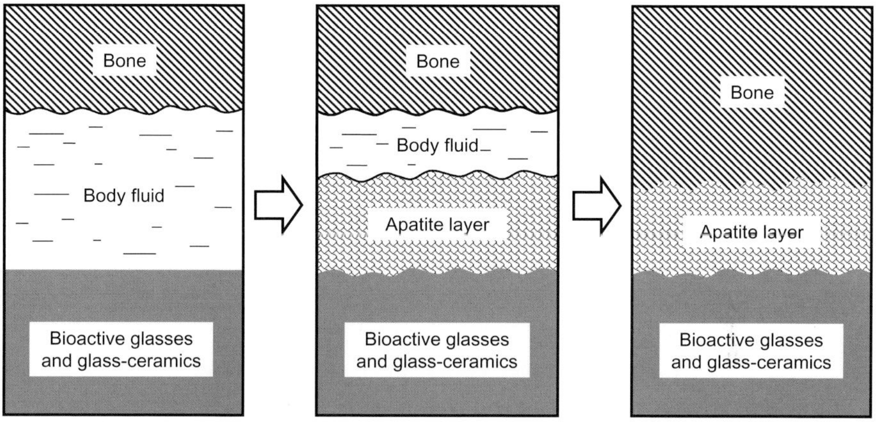


Figure 8.2 Schematic illustration of the bone-bonding process of bioactive glasses and glass-ceramics to living bone. In a bony defect, bioactive glass materials bond to bone following coating of the materials with apatite layer.

Table 8.1 Inorganic ion concentrations of human plasma and the SBF developed by Kokubo *et al.*

| | Concentration (mol m$^{-3}$) | |
|---|---|---|
| Ion | SBF | Human plasma |
| $Na^+$ | 142.0 | 142.0 |
| $K^+$ | 5.0 | 5.0 |
| $Mg^{2+}$ | 1.5 | 1.5 |
| $Ca^{2+}$ | 2.5 | 2.5 |
| $Cl^-$ | 147.8 | 103.0 |
| $HCO_3^-$ | 4.2 | 27.0 |
| $HPO_4^{2-}$ | 1.0 | 1.0 |
| $SO_4^{2-}$ | 0.5 | 0.5 |

does not involve any special techniques and it can be stored for a maximum of 30 days[6] and used easily. Therefore, SBF is widely used towards the development of bioactive materials.

The hydroxyapatite formed on the surface of bioactive glasses and glass-ceramics shows characteristics of carbonate-containing hydroxyapatite with small crystallites and/or a defective structure.[7]

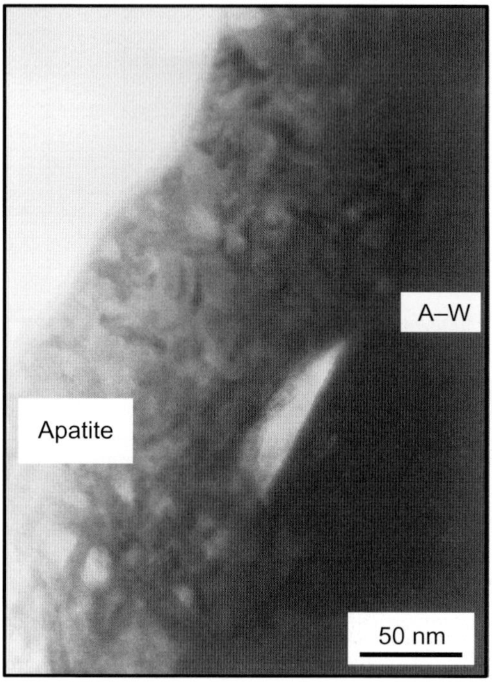


Figure 8.3 TEM image of apatite formed on glass-ceramic containing oxyfluorapatite and $\beta$-wollastonite after soaking in SBF for 30 days. A–W: glass-ceramic containing oxyfluorapatite and $\beta$-wollastonite. Reproduced with permission from Ref. 8. Copyright 1995, Ceramic Society of Japan.

The characteristics of the hydroxyapatite are similar to those displayed by natural bone. Accordingly, the hydroxyapatite is typically referred as "bone-like apatite". Figure 8.3[8] shows the bone-like apatite layer formed on the bioactive glass-ceramic containing oxyfluorapatite and $\beta$-wollastonite after incubation in SBF. Because the bone-like apatite layer forms after exposure to the body environment, the osteoblast cells would grow preferentially to fibroblast cells, which produce fibrous tissues on the material surface. This preferential growth enables direct contact with the bone without the intervention of fibrous tissues. In contrast, alumina ($Al_2O_3$) and zirconia ($ZrO_2$) ceramics, which are representative bioinert ceramics, are encapsulated by fibrous tissues and do not make direct contact to bone. SBF can mimic bone-like apatite formation on the material surface after implantation in bony defects though SBF does not contain any biological entities such as cells and proteins. In other words, bone-like apatite formation can occur in the absence of biological entities. This observation implies that bone-like apatite formation on the surface of ceramic materials is related to inorganic reactions between the ceramic material and surrounding solution. The SBF developed by Kokubo *et al.* can be effectively used as a screening test for developing novel bioactive materials showing bone-bonding properties based on the ability to form bone-like apatite.

Bone-like apatite formation on the surface of materials strongly depends on the composition of the substrate. The composition dependency of apatite formation on the surface of $CaO–SiO_2–P_2O_5$ glass in SBF is shown in Figure 8.4.[9] Glass materials that are mainly composed of CaO and $SiO_2$ displayed bone-like apatite-forming ability on their surface in SBF. In contrast, glass materials that are mainly composed of CaO and $P_2O_5$ do not display such capability. This result implies that glass materials that are mainly composed of CaO and $SiO_2$ have high potential to featuring bone-bonding properties. It should be noted that glass materials in binary $CaO–SiO_2$

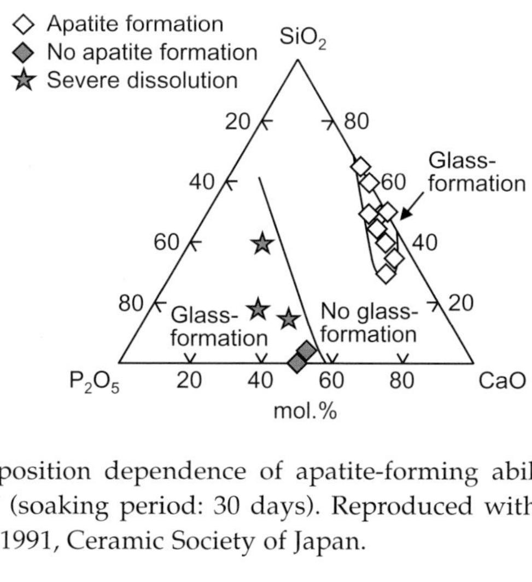


Figure 8.4 Composition dependence of apatite-forming ability of $CaO–SiO_2–P_2O_5$ glass in SBF (soaking period: 30 days). Reproduced with permission from Ref. 9. Copyright 1991, Ceramic Society of Japan.

system also show potential of bone-like apatite formation in SBF. This finding indicates that surface reactions between substrates composed of CaO and $SiO_2$ and the surrounding fluid induce bone-like apatite formation in the body environment. The formation mechanism of bone-like apatite on substrates is well understood as follows.[10] Dissolution of $Ca^{2+}$ from a CaO–$SiO_2$ glass substrate leads to an increased degree of supersaturation of the surrounding fluid with respect to hydroxyapatite. Also, the formation of silica hydrogel rich in silanol groups (Si–OH) is favoured, which induces the heterogeneous nucleation of bone-like apatite. The presence of Si–OH on the surface of CaO–$SiO_2$ glass is responsible for the higher bone-like apatite-forming ability of CaO–$SiO_2$ glass when compared with that of CaO–$P_2O_5$ glass. Therefore, CaO and $SiO_2$ are often integrated as the main components of bioactive glass and bioactive glass-ceramic materials.

## 8.3 Bioactive Glass-Ceramics for Bone Repair

A typical example of bioactive glass termed 45S5 Bioglass® has a nominal composition of 45 wt.% $SiO_2$, 24.5 wt.% $Na_2O$, 24.5 wt.% CaO, and 6 wt.% $P_2O_5$.[10] 45S5 glass displays high biocompatibility towards both hard and soft tissues. However, this glass material does not feature satisfactory mechanical properties for repairing hard tissues under high load. Accordingly, the glass material is applied for repairing auditory ossicles and filling residual ridge. Crystallisation of glass is effective for improving the mechanical properties of bone-repairing materials. Bioactive glass-ceramics have also been developed, and some of them have been used clinically.

### 8.3.1 Representative bioactive glass-ceramics

Ceravital®, Cerabone®A–W, and Bioverit® are representative bioactive glass-ceramics developed as bone-repairing biomaterials. Their phase and composition are summarised in Table 8.2 and compared with those of Bioglass®.[11] Ceravital® is a type of glass-ceramic that is only composed of apatite (oxyapatite; $Ca_{10}(PO_4)_6O$) as the

Table 8.2 Phase and composition of Bioglass® (45S5) and representative bioactive glass-ceramics.[11]

| | Bioglass® (45S5) | Ceravital® | Cerabone®A–W | Bioverit® |
|---|---|---|---|---|
| Phase | Glass | Apatite, glass | Apatite, $\beta$-wollastonite, glass | Apatite, phlogopite, glass |
| Compositions/ mass% | | | | |
| $Na_2O$ | 24.5 | 5–10 | 0 | 3–8 |
| $K_2O$ | 0 | 0.5–3.0 | 0 | 3–8 |
| MgO | 0 | 2.5–5.0 | 4.6 | 2–21 |
| CaO | 24.5 | 30–35 | 44.7 | 10–34 |
| $Al_2O_3$ | 0 | 0 | 0 | 8–15 |
| $SiO_2$ | 45.0 | 40–50 | 34.0 | 19–54 |
| $P_2O_5$ | 6.0 | 10–50 | 16.2 | 2–10 |
| $CaF_2$ | 0 | | 0.5 | 3–23 |
| $B_2O_3$ | 0 | | | |

crystalline phase. Ceravital® was developed by Brömer *et al.*[12] in 1973. Its main components are $Na_2O$, CaO, $SiO_2$, and $P_2O_5$, similar to those present in Bioglass®. The alkaline component amount of Ceravital® is lower than that of Bioglass®. Consequently, Ceravital® displays higher chemical durability than Bioglass® in *in vivo* environment. In contrast, Cerabone®A–W is composed of apatite (oxyfluoroapatite), $\beta$-wollastonite and glass. This glass-ceramic material was developed by Kokubo and is often called glass-ceramic A–W.[13] Glass-ceramic A–W is prepared as follows.[14] A mixture of MgO, CaO, $SiO_2$, $P_2O_5$, and $CaF_2$ at a mass ratio of 4.6:44.9:34.2:16.3:0.5 is melted in a platinum crucible at 1450°C. The melt is then quenched to obtain a glass material. Then the glass is pulverised, and the resulting glass powder is heated to 1050°C for sintering and crystallisation. Upon heating, apatite and $\beta$-wollastonite precipitate in the glass matrix. Figure 8.5[8] shows the microstructure of the obtained glass-ceramic A–W. Granules of apatite and $\beta$-wollastonite with a

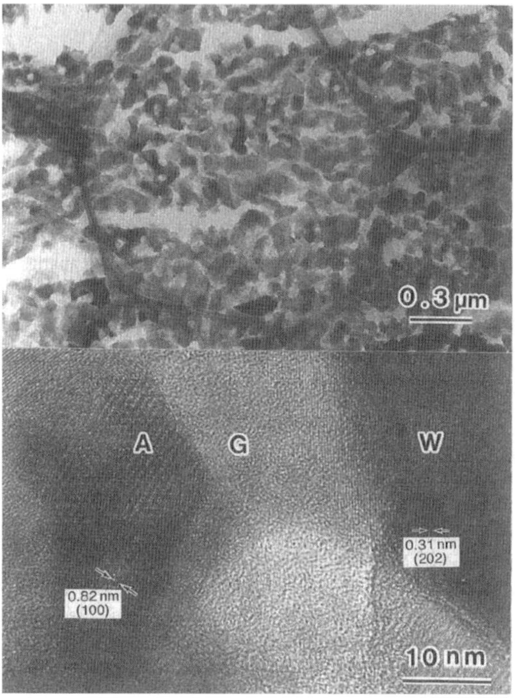


Figure 8.5 Low- and high-magnification TEM images of glass-ceramic A–W. A: apatite; W: β-wollastonite; G: glassy phase. Reproduced with permission from Ref. 8. Copyright 1995, Ceramic Society of Japan.

size of 50–100 nm can be observed. Additionally, the granules are homogeneously dispersed in the glass matrix. Bioverit® is a machinable glass-ceramic that was developed by Höland *et al.* Bioverit® is composed of apatite (oxyapatite) and phlogopite ($(Na,K)Mg_3[AlSi_3O_{10}]F_2$).[15–17] Figure 8.6[15] shows the microstructure of Bioverit®-type glass-ceramic. Large crystals in the order of a few micrometres in size and small crystals in the order of submicrometres in size, corresponding to phlogopite and apatite, respectively, can be observed. Phlogopite crystals dispersed in the glass matrix inhibits crack extension during cutting work, thereby bestowing the glass-ceramic Bioverit® with high machinability.

Improving the mechanical strength of glass materials is the main motivation for crystallising glass materials towards the

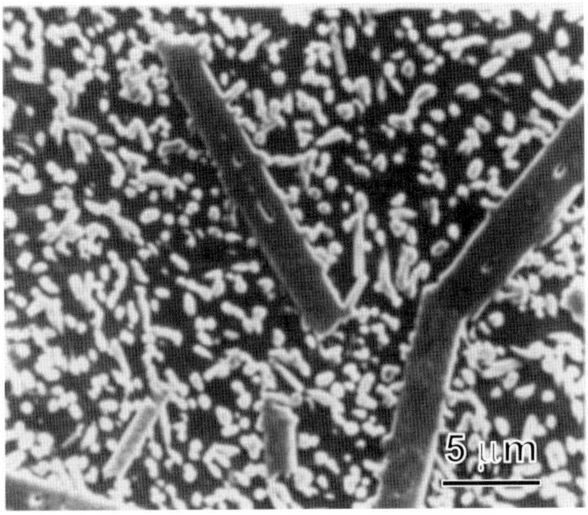


Figure 8.6 Scanning electron microscope (SEM) image of glass-ceramic containing apatite and phlogopite. Reproduced with permission from Ref. 15. Copyright 1985, Wiley-VCH Verlag GmbH & Co. KGaA.

development of biomaterials. Table 8.3 summarises the mechanical properties of Bioglass®, crystallised bioactive glass-ceramics, and two types of natural bones.[10] As observed, the mechanical strength, Young's modulus, and Vickers hardness of the crystalline bioactive glass-ceramics are higher than those of Bioglass® owing to crystal precipitation in the glass matrix. Specifically, Cerabone®A–W displays a bending strength of 215 MPa, which is higher than that displayed by Bioglass®. Moreover, the bending strength of Cerabone®A–W is superior to that of cortical bone. Thus, Cerabone®A–W is a better material in terms of mechanical properties. The high strength properties of bioactive glass-ceramics contribute to the extended lifetime of the materials. The mechanical strength of a material typically decreases owing to fatigue in *in vivo* environment under applied stress. Therefore, evaluating the fatigue properties of a target material is essential for application as artificial bone. Figure 8.7[18] shows the relationship between the lifetime of glass and different glass-ceramic materials in SBF as a function of applied stress. Under an applied stress of 65 MPa, it is estimated that a sintered hydroxyapatite ceramic would fracture within 1 min. In contrast, glass-ceramic

Table 8.3 Mechanical properties of bioactive glass materials and bones.[11]

| | Bioglass® | Ceravital® | Cerabone® A–W | Bioverit® | Cortical bone | Cancellous bone |
|---|---|---|---|---|---|---|
| Compressive strength (MPa) | | 500 | 1080 | 500 | | 2–12 |
| Bending strength (MPa) | 42* | | 215 | 100–160 | 50–150 | |
| Young's modulus (GPa) | 35 | 100–150 | 218 | 70–88 | 7–30 | 0.05–0.5 |
| Fracture toughness (MPa $m^{1/2}$) | | | 2.0 | 0.5–1.0 | 2–12 | |
| Vickers hardness | 458 ± 9.4 | | 680 | 500 | | |

*Tensile strength.

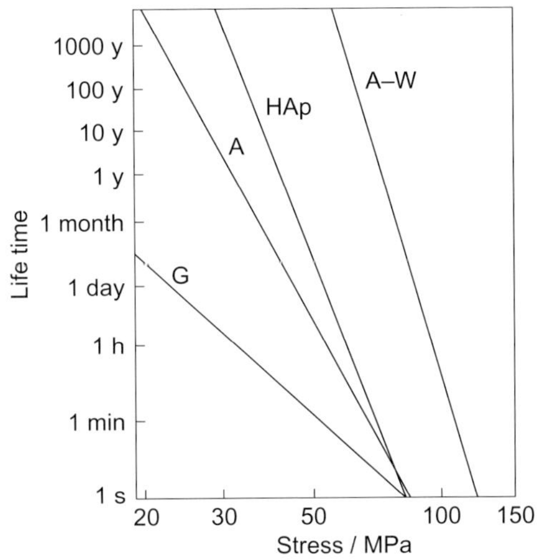


Figure 8.7 Changes in the lifetime of different materials in SBF as a function of applied stress. G: glass; A: glass-ceramic containing apatite; A–W: glass-ceramic containing apatite and $\beta$-wollastonite; HAp: sintered apatite ceramic. Reproduced with permission from Ref. 18. Copyright 1987, Springer.

A–W is estimated to have a prolonged lifetime of 10 years before breakage under similar conditions. Breakage of bone-repairing materials after implantation poses a large burden on the patient. Improved mechanical performances, including fatigue characteristics, contribute to the material reliability.

Bioactive glass-ceramics were applied clinically (Figure 8.8).[16,19–21] They contributed towards improving the quality of life of patients with bony defects. However, the mechanical, physicochemical, and biological performances of glass-ceramic artificial bones are not as efficient as those of autologous bone. Hence, continuing research and development of new glass-ceramic artificial bones is important. In the next section, the recent progress achieved in bioactive glass-ceramics is presented.

### 8.3.2 Recent progress in bioactive glass-ceramics

Bioactive glass-ceramics for bone repair have been designed to display high biological affinity for living bone as well as

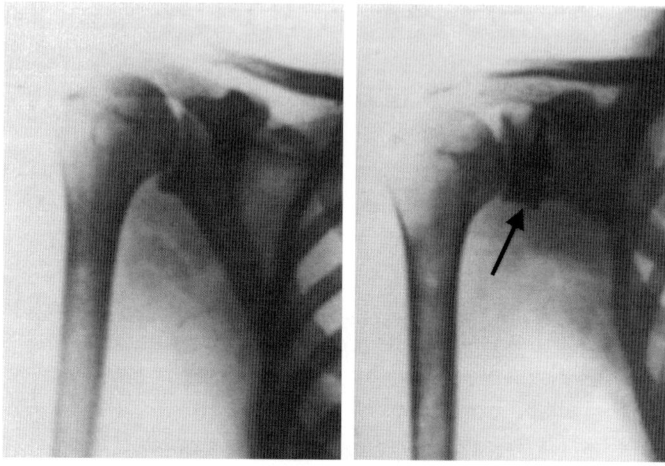

(a) (b)

Figure 8.8 Application of a wedge-shaped implant made of bioactive glass-ceramic for treating recurrent dislocation of the shoulder (a) before and (b) after surgery. Reproduced with permission from Ref. 16. Copyright 1987, Wiley-VCH Verlag GmbH & Co. KGaA.

mechanical properties comparable to those displayed by natural cortical bone. As the mechanism of bone bonding was elucidated in past studies, the subsequent investigations on the improvement of mechanical strength of bioactive glass-ceramics were conducted in terms of composition, synthesis procedures, and material morphology. These studies are discussed below.

The mechanical properties of $MgO–CaO–SiO_2–P_2O_5–CaF_2$ glass-ceramic were improved by $ZrO_2$ addition.[22] This glass-ceramic was composed of fluorapatite, $\beta$-wollastonite, and monoclinic-$ZrO_2$ or tetragonal-$ZrO_2$. Tetragonal-$ZrO_2$ increased the fracture toughness, bending strength, and micro-hardness of the glass-ceramic, whereas monoctinic-$ZrO_2$ did not exert such effects. $ZrO_2$ addition also improved the mechanical properties, apatite-forming ability in SBF, and cell proliferation (MG63 cell line) in $Na_2O–CaO–SiO_2–P_2O_5$ glass-ceramic.[23]

Incorporating Zn in ceramic biomaterials is also an attractive method for improving the biological activity of bone re-modelling because released $Zn^{2+}$ can act as an activator in bone metabolism. Zn is an essential element that has stimulatory effects on bone formation.[24] Zn released from glass-ceramics is expected to show novel effects, namely enhancement of bone regeneration. The effects of ZnO addition in $CaO–SiO_2–P_2O_5–CaF_2$ glass on the physicochemical property of the obtained glass-ceramic were investigated.[25] The heat-treated glass material with a nominal composition of $x$ZnO–(57−$x$)CaO–35.4$SiO_2$–7.2$P_2O_5$–0.4$CaF_2$ ($x$ = 0–14.2) in molar ratio was composed of apatite and $\beta$-wollastonite. The chemical durability of the glass-ceramic increased with increasingly added ZnO contents. While, its apatite-forming ability decreased owing to the decreasing reactivity of the glass-ceramic upon ZnO addition. In the case of $Na_2O–MgO–CaO–SiO_2–P_2O_5–CaF_2$ glass-ceramic, ZnO addition also resulted in a reduced apatite-forming ability of the glass-ceramic. In contrast, $TiO_2$ addition in this compositional system promoted apatite formation.[26]

Improvements in the mechanical properties were additionally attempted by developing glass-ceramics containing calcium pyrophosphate crystals.[27] The developed glass-ceramic $SiO_2$–($Al_2O_3$)–MgO–$Na_2O$–$K_2O$–F–CaO–$P_2O_5$ displayed machinability. The effect of dopants on the structure and crystallisation behaviour of the glass was studied by various techniques including X-ray diffraction, differential scanning calorimetry, and solid-state nuclear magnetic resonance spectroscopy.[28–30] Furthermore, glass-ceramics containing various crystalline phases with wide composition ranges can be prepared. This composition flexibility enables the development of bioactive materials and broadening of biomedical applications. The preparation of glass-ceramics with diverse compositions is possible owing to the availability of different routes for the preparation of glass. For instance, the application of sol–gel methods affords the preparation of bioactive glass materials with wider compositions when compared with the conventional melt-quenching route. Silicate-type[31–33] and phosphate-type glass-ceramics[34] were synthesised by sol–gel methods; these glass-ceramics displayed various mechanical and biological properties.

For bone tissue repair applications, porous structures are attractive for bone ingrowth after implantation and as scaffolds for tissue regeneration. A type of porous glass-ceramic was prepared by sintering and crystallisation of Bioglass® powder at 1000°C.[35] The prepared material exhibited osteoconduction as well as osteoinduction capabilities. Bioactive porous glass-ceramics with interconnected pores can be fabricated by sintering glass powder mounted on an organic polymer template such as sponge materials. Typically, the sponge material was dipped in the glass slurry and dried. Then the specimen was heat-treated through appropriate temperature profiles to remove the organic template and to subsequently sinter and crystallise the glass matrix.[36–38] Figure 8.9 shows the structure of a porous glass-ceramic material.[39] To improve the mechanical strength of the porous glass-ceramic, a polymer coating was applied on the framework of the porous material.[40] Additionally, microspheres, in the size range of 75–150 $\mu$m, of bioactive glass-ceramics have been developed as fillers for bone defects owing to their ease of handling and controlled reactivity in surrounding body fluid. Fu *et al.*[41] prepared such microspheres by spheroidisation of 45S5 glass powder in a flame. The authors conducted a theoretical investigation on the long-term reaction (10 years) of the glass-ceramic in phosphate aqueous solutions. The crystalline

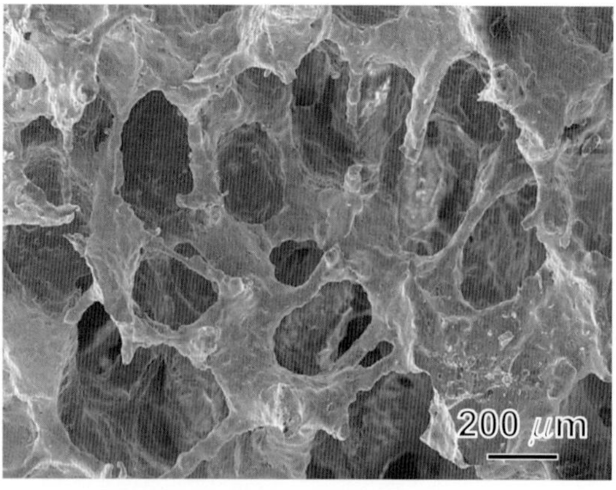


Figure 8.9 SEM image of porous glass-ceramic. Reproduced with permission from Ref. 39. Copyright 2007, Elsevier.

phase of the deposits generated on the micro-spheres was believed to be bone-like apatite, of which the characteristics were affected by pH and phosphate concentration of the solution. For instance, apatite formation was observed when the pH of the phosphate solution was 9.5 irrespective of the phosphate concentration. In contrast, when the pH of the phosphate solution was 7, calcium potassium pyrophosphate at a high concentration formed instead of apatite. The results offered a fundamental model for predicting the long-term conversion behaviour of the bioactive glass-ceramic in body environment.

Additionally, bioactive glass-ceramics have been used as fillers to prepare composites with polymeric materials. Poly(methyl methacrylate) (PMMA) was used as bone cement for fixation of artificial hip joints. PMMA is commonly used in orthopaedic surgery though it is bioresistant to encapsulation with fibrous tissues after implantation in bone. Composites of PMMA and bioactive glass-ceramics have been investigated to produce materials with bone-bonding capabilities and suitable mechanical properties comparable to those of bone.[42,43] The PMMA/glass-ceramic A–W composite bonded to bone directly though PMMA cement did not bond to bone (Figure 8.10).[44] The compressive and bending strength of the composite were 125 and 116 MPa, respectively. These values were higher than those displayed by PMMA cement.

## 8.4 Application of Glass-Ceramics in Medicine Other Than Bone Repair

The biomedical application of glass-ceramics does not only relate to bone repair though studies on bone-bonding properties have mainly focused on the glass-ceramics described above. The bioactive glass-ceramics mentioned above exhibit reactivity with the surrounding body fluid and thus display high biological affinity for living bone. Appropriate selection of the glass composition allows for the formation of crystals in glass-ceramics with various functions. In this section, the applications of glass-ceramics in cancer and dental therapy are briefly discussed.

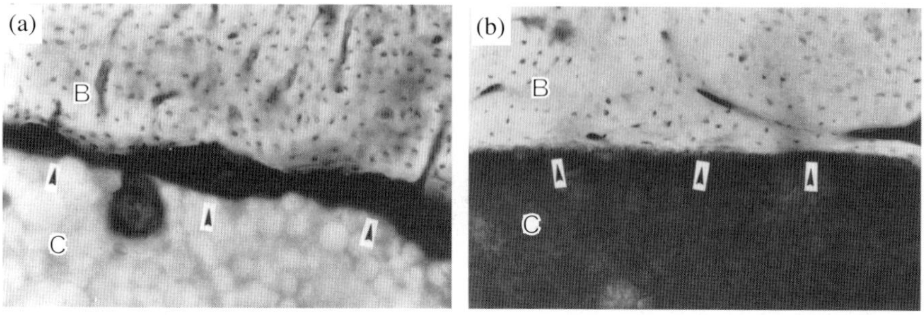


Figure 8.10 Microphotographs showing axial sections of rat tibiae implanted with (a) PMMA cement and (b) PMMA/glass-ceramic A–W composite for 8 weeks. Reproduced with permission from Ref. 44. Copyright 2000, Elsevier.

### 8.4.1 Cancer therapy

Hyperthermia, a type of cancer treatment, has received much attention because it is a low-invasive treatment. Particularly, hyperthermia involving magnetic glass-ceramics is regarded as a safe method that is used to treat tumours located deep within the body such as the bone and liver. Bioactive glass-ceramics containing magnetite were investigated for bone tumour treatment.[45,46] Additionally, glass-ceramics containing various types of ferrites, such as Li-ferrite[47], $\alpha$-Fe,[48] Zn-ferrite,[49] Li, Mn-ferrite,[50] and Mg-ferrite,[51] were developed, and some of them exhibit bioactivity. The incorporation of ferrimagnetic bioactive glass-ceramic into PMMA bone cement was also attempted.[52] This type of glass-ceramic featured both bone-repairing and heat-generation properties.

### 8.4.2 Dental therapy

Glass-ceramics have also been applied in dental therapy.[53] Biomaterials used in the restoration of teeth do not have to fulfil any bioactive functions. However, they must demonstrate excellent mechanical properties and chemical durability in the oral cavity, and optical properties that are comparable with those displayed by natural teeth, as reviewed by Höland and Beall.[54] Mica-based glass-ceramics, leucite-based glass-ceramics, and lithium disilicate glass-ceramics have been clinically used. Artificial crowns are typical examples of the application of glass-ceramics. The colour tone, translucency, and thermal conductivity of glass-ceramics can be controlled by adjusting the composition and crystallisation conditions. Glass-ceramics containing apatite and magnesium titanate[55,56] as well as lithium disilicate[57,58] have been developed as materials that have properties suitable for natural tooth crown.

In dentistry, both the antibacterial property and bone-bonding property are useful under oral conditions. Bioactive glass-ceramics with antibacterial property have been studied as bone-repairing materials[59]; however, some glass-ceramics with antibacterial property are also suitable for dental applications.[60–62]

## 8.5 Conclusions

In this chapter, glass-ceramics with specific biological activity for biomaterial applications were reviewed. The mechanism of bone bonding of bioactive glass ceramics was first explained. The formation of bone-like apatite on glass-ceramic surface is essential for direct bonding of glass-ceramics to living bone. The bone-like apatite-forming ability of glass-ceramics under *in vivo* conditions can be probed in SBF. Subsequently, the properties of clinically applied bioactive glass-ceramics, i.e., Ceravital®, Cerabone®A–W, and Bioverit®, were introduced and compared to those of natural bone. It is worth noting that Cerabone®A–W displayed higher mechanical strength than cortical bone. Recent progress in the development of bioactive glass-ceramics was then presented. Many research studies have focused on improving the biological and mechanical properties of glass-ceramics by controlling their composition and microstructures as well as by fabricating hybrids with polymers. Finally, the application of glass-ceramics in other fields, i.e., cancer and dental therapy, was briefly presented. Glass-ceramics have been one of the most useful biomedical materials developed since the early development of bioactive ceramic materials and they are presently generating much interest in the bioceramic field.

## 8.6 Future Perspectives

Glass-ceramics are composites constituting of crystal and glass phases. The crystal phase, crystal size, composition of the glass phase, and ratio of the crystal and glass phases can be controlled. Accordingly, the physicochemical, mechanical, and biological properties of glass-ceramics can be tailored to meet the requirements of different applications. Glass-ceramics have been mainly used in the reconstruction of hard tissues of bone and teeth. Additionally, they have high potential in other medical applications as various functionalities can be exhibited by tailoring the nature and composition of the crystalline phase precipitated in the glass matrix.

## References

1. Hench LL, Splinter RJ, Allen WC, Greenlee TK. Bonding mechanisms at the interface of ceramic prosthetic materials. *J Biomed Mater Res Symp* 1971;2:117–141.
2. Kokubo T. Surface chemistry of bioactive glass-ceramics. *J Non-Cryst Solids* 1990;120:138–151.
3. Kokubo T. Bioactive glass ceramics: Properties and applications. *Biomaterials* 1991;12:155–163.
4. Neo M, Kotani S, Nakamura T, Yamamuro T, Ohtsuki C, Kokubo T, Bando Y. A comparative study of ultrastructures of the interfaces between four kinds of surface-active ceramic and bone. *J Biomed Mater Res* 1992;26:1419–1432.
5. Kokubo T, Kushitani H, Sakka S, Kitsugi T, Yamamuro T. Solutions able to reproduce *in vivo* surface-structure changes in bioactive glass-ceramic A–W. *J Biomed Mater Res* 1990;24:721–734.
6. ISO 23317:2014(en), Implants for surgery — *In vitro* evaluation for apatite forming ability of implant materials.
7. Kokubo T. A/W glass-ceramic: Processing and properties. In: Hench LL, Wilson J, editors. *An Introduction to Bioceramics*. Singapore: World Scientific; 1993; pp. 75–88.
8. Ohtsuki C, Aoki Y, Kokubo T, Bando Y, Neo M, Nakamura T. Transmission electron microscopic observation of glass-ceramic A–W and apatite layer formed on its surface in a simulated body fluid. *J Ceram Soc Jpn* 1995;103:449–454.

9. Ohtsuki C, Kokubo T, Takatsuka K, Yamamuro T. Compositional dependence of bioactivity of glasses in the system $CaO–SiO_2–P_2O_5$: Its *in vitro* evaluation. *J Ceram Soc Jpn* 1991;99:1–6.
10. Ohtsuki C, Kokubo T, Yamamuro T. Mechanism of apatite formation on $CaO–SiO_2–P_2O_5$ glasses in a simulated body fluid. *J Non-Cryst Solids* 1992;143:84–92.
11. Hench LL. Bioceramics. *J Am Ceram Soc* 1998;81:1705–1728.
12. Brömer H, Pfeil E, Käs HH. German Patent No. 2,326,100; 1973.
13. Kokubo T, Shigematsu M, Nagashima Y, Tashiro M, Nakamura T, Yamamuro T, Higashi S. Apatite- and wollastonite-containing glass-ceramics for prosthetic application. *Bull Inst Chem Res Kyoto Univ* 1982;60:260–268.
14. Kokubo T, Ito S, Sakka S, Yamamuro T. Formation of a high-strength bioactive glass-ceramic in the system $MgO–CaO–SiO_2–P_2O_5$. *J Mater Sci* 1986;21:536–540.
15. Höland W, Vogel W, Naumann K. Interface reactions between machinable bioactive glass-ceramics and bone. *J Biomed Mater Res* 1985;19:303–312.
16. Vogel W, Höland W. The development of bioglass ceramics for medical applications. *Angew Chem Int Ed Engl* 1987;26:527–544.
17. Höland W, Wange P, Naumann K, Vogel J, Carl G, Jana C, Götz W. Control of phase formation processes in glass-ceramics for medicine and technology. *J Non-Cryst Solids* 1991;129:152–162.
18. Kokubo T, Ito S, Shigematsu M, Sakka S, Yamamuro T. Fatigue and life-time of bioactive glass-ceramic A–W containing apatite and wollastonite. *J Mater Sci* 1987;22:4067–4070.
19. Yamamuro T. A/W glass-ceramic: Clinical application. In Hench LL, Wilson J, editors. *An Introduction to Bioceramics*. Singapore: World Scientific; 1993; pp. 89–103.
20. Yamamuro T. Clinical application of bioactive glass-ceramics. In Kokubo T, editor. *Bioceramics and Their Clinical Applications*. USA: CRC Press; 2008; pp. 583–605.
21. Crovace MC, Souza MT, Chinaglia CR, Peitl O, Zanotto ED. Biosilicate® — A multipurpose, highly bioactive glass-ceramic. *In vitro, in vivo* and clinical trials. *J Non-Cryst Solids* 2015; published online, doi:10.1016/j.jnoncrysol.2015.03.022.
22. Li HC, Wang DG, Meng XG, Chen CZ. Effect of $ZrO_2$ additions on the crystallization, mechanical and biological properties of $MgO–CaO–SiO_2–P_2O_5–CaF_2$ bioactive glass-ceramics. *Colloid Surf B* 2014;118:226–233.

23. Mondal D, So-Ra S, Lee BT. Fabrication and characterization of $ZrO_2$–CaO–$P_2O_5$–$Na_2O$–$SiO_2$ bioactive glass ceramics. *J Mater Sci* 2013;48: 1863–1872.
24. Yamaguchi M, Oishi H, Suketa Y. Stimulatory effect of zinc on bone formation in tissue culture. *Biochem Pharmacol* 1987;36:4007–4012.
25. Kamitakahara M, Ohtsuki C, Inada H, Tanihara M, Miyazaki T. Effect of ZnO addition on bioactive CaO–$SiO_2$–$P_2O_5$–$CaF_2$ glass-ceramics containing apatite and wollastonite. *Acta Biomater* 2006;2: 467–471.
26. Salman SM, Salama SN, Darwish H, Abo-Mosallam HA. *In vitro* bioactivity of glass-ceramics of the $CaMgSi_2O_6$–$CaSiO_3$–$Ca_5(PO_4)_3F$–$Na_2SiO_3$ system with $TiO_2$ or ZnO additives. *Ceram Int* 2009;35:1083–1093.
27. Kasuga T, Terada M, Nogami M. Machinable calcium pyrophosphate glass-ceramics. *J Mater Res* 2001;16:876–880.
28. Hill RG, Stamboulis A, Law RV, Clifford A, Towler MR, Crowley C. The influence of strontium substitution in fluorapatite glasses and glass-ceramics. *J Non-Cryst Solids* 2004;336:223–229.
29. Calver A, Hill RG, Stamboulis A. Influence of fluorine content on the crystallization behavior of apatite–wollastonite glass-ceramics. *J Mater Sci* 2004;39:2601–2603.
30. Kansal I, Tulyaganov DU, Goel A, Pascual MJ, Ferreira JMF. Structural analysis and thermal behavior of diopside–fluorapatite–wollastonite-based glasses and glass-ceramics. *Acta Biomater* 2010;6:4380–4388.
31. Román J, Padilla S, Vallet-Regí M. Sol–gel glasses as precursors of bioactive glass ceramics. *Chem Mater* 2003;15:798–806.
32. Bini M, Grandi S, Capsoni D, Mustarelli P, Saino E, Visai L. $SiO_2$–$P_2O_5$–CaO glasses and glass-ceramics with and without ZnO: Relationships among composition, microstructure, and bioactivity. *J Phys Chem* 2009;113:8821–8828.
33. Adams LA, Essien ER, Shaibu RO, Oki A. Sol–gel synthesis of $SiO_2$–CaO–$Na_2O$–$P_2O_5$ bioactive glass ceramic from sodium metasilicate. *New J Glass Ceram* 2013;3:11–15.
34. Cai S, Li J, Xu G, Li X, Ye X, Jiang W. *In vitro* solubility and bioactivity of Sr and Mg co-doped calcium phosphate glass-ceramics derived from different heat-treatment temperatures. *Mater Chem Phys* 2011;131:462–470.
35. Yuan H, de Bruijn JD, Zhang X, van Blitterswijk CA, de Groot K. Bone induction by porous glass ceramic made from Bioglass® (45S5). *J Biomed Mater Res* 2001;58:270–276.

36. Ohtsuki C, Miyazaki T, Kishiro K, Kamitakahara M, Tanihara M. Bioactivity of MgO–CaO–$SiO_2$ porous glass-ceramics. *J Ceram Soc Jpn* 2004;112:S809–S812.
37. Kamitakahara M, Ohtsuki C, Kozaka Y, Ogata S, Tanihara M, Miyazaki T. Preparation of porous glass-ceramics containing whitlockite and diopside for bone repair. *J Ceram Soc Jpn* 2006;114:82–86.
38. Chen QZ, Thompson ID, Boccaccini AR. 45S5 Bioglass®-derived glass-ceramic scaffolds for bone tissue engineering. *Biomaterials* 2006;27:2414–2425.
39. Vitale-Brovarone C, Verné E, Robiglio L, Appendino P, Bassi F, Martinasso G, Muzio G, Canuto R. Development of glass-ceramic scaffolds for bone tissue engineering: Characterisation, proliferation of human osteoblasts and nodule formation. *Acta Biomater* 2007;3: 199–208.
40. Bretcanu O, Misra S, Roy I, Renghini C, Fiori F, Boccaccini AR, Salih V. *In vitro* biocompatibility of 45S5 Bioglass®-derived glass ceramic scaffolds coated with poly(3-hydroxybutyrate). *J Tissue Eng Regen Med* 2009;3:139–148.
41. Fu H, Rahaman MN, Day DE, Huang W. Long-term conversion of 45S5 bioactive glass-ceramic microspheres in aqueous phosphate solution. *J Mater Sci Mater Med* 2012;23:1181–1191.
42. Mousa WF, Kobayashi M, Kitamura Y, Zeineldin IA, Nakamura T. Effect of silane treatment and different resin compositions on biological properties of bioactive bone cement containing apatite–wollastonite glass ceramic powder. *J Biomed Mater Res* 1999;47:336–344.
43. Shinzato S, Kobayashi M, Mousa WF, Kamimura M, Neo M, Kitamura Y, Kokubo T, Nakamura T. Bioactive polymethyl methacrylate-based bone cement: Composition of glass beads, apatite- and wollastonite-containing glass-ceramic, and hydroxyapatite fillers on mechanical and biological properties. *J Biomed Mater Res* 2000;51:258–272.
44. Mousa WF, Kobayashi M, Shinzato S, Kamimura M, Neo M, Yoshihara S, Nakamura T. Biological and mechanical properties of PMMA-based bioactive bone cements. *Biomaterials* 2000;21:2137–2146.
45. Ebisawa Y, Miyaji F, Kokubo T, Ohura K, Nakamura T. Bioactivity of ferrimagnetic glass-ceramics in the system FeO–$Fe_2O_3$–CaO–$SiO_2$. *Biomaterials* 1997;18:1277–1284.
46. Kawashita M, Takaoka H, Kokubo T, Yao T, Hamada S, Shinjo T. Preparation of magnetite-containing glass-ceramics in controlled atmosphere for hyperthermia of cancer. *J Ceram Soc Jpn* 2001;109:39–44.

47. Luderer AA, Borrelli NF, Panzarino JN, Mansfield GR, Hess DM, Brown JL, Barnett EH, Hahn EW. Glass-ceramic-mediated, magnetic-field-induced localized hyperthermia: Response of a murine mammary carcinoma. *Radiat Res* 1983;94:190–198.
48. Konaka H, Miyaji F, Kokubo T. Preparation and magnetic properties of glass-ceramics containing α-Fe for hyperthermia. *J Ceram Soc Jpn* 1997;105:833–836.
49. Kawashita M, Iwahashi Y, Kokubo T, Yao T, Hamada S, Shinjo T. Preparation of glass-ceramics containing ferrimagnetic zinc–iron ferrite for the hyperthermal treatment of cancer. *J Ceram Soc Jpn* 2004;112:373–379.
50. Hsi CS, Cheng HZ, Hsu HJ, Chen YS, Wang MC. Crystallization kinetics and magnetic properties of iron oxide contained $25Li_2O$–$8MnO_2$–$20CaO$–$2P_2O_5$–$45SiO_2$ glasses. *J Eur Ceram Soc* 2007;27: 3171–3176.
51. Li GD, Zhou DL, Lin Y, Pan TH, Chen GS, Yin QD. Synthesis and characterization of magnetic bioactive glass-ceramics containing Mg ferrite for hyperthermia. *Mater Sci Eng C* 2010;30:148–153.
52. Bruno M, Miola M, Bretcanu O, Vitale-Brovarone C, Gerbaldo R, Laviano F, Verné E. Composite bone cements loaded with a bioactive and ferrimagnetic glass-ceramic. Part I: morphological, mechanical and calorimetric characterization. *J Biomater Appl* 2014;29:254–267.
53. Höland W, Rheinberger V. Dental glass-ceramics. In Kokubo T, editor. *Bioceramics and their Clinical Applications*. USA: CRC Press; 2008; pp. 548–568.
54. Höland W, Beall GH. *Glass-Ceramic Technology*. USA: Wiley; 2012; pp. 315–342.
55. Kokubo T, Sakka S, Sako M, Ikejiri S. Preparation of glass-ceramic containing crystalline apatite and magnesium titanate for dental crowns. *J Ceram Soc Jpn* 1989;97:239–244.
56. Tada Y, Kawano F, Kon M, Matsumoto N, Asaoka K. Influence of crystallization on strength and color of castable glass-ceramics containing two crystals. *Bio-Med Mater Eng* 1995;5:233–243.
57. Chaiyabutr Y, Kois JC, LeBeau D, Nunokawa G. Effect of abutment tooth color, cement color, and ceramic thickness on the resulting optical color of a CAD/CAM glass-ceramic lithium disilicate-reinforced crown. *J Prosthet Dent* 2011;105:83–90.
58. Zhao K, Wei YR, Pan Y, Zhang XP, Swain MV, Guess PC. Influence of veneer and cyclic loading on failure behavior of lithium disilicate glass-ceramic molar crowns. *Dent Mater* 2014;30:164–171.

59. Vitale-Brovarone C, Miola M, Balagna C, Verné E. 3D-glass-ceramic scaffolds with antibacterial properties for bone grafting. *Chem Eng J* 2008;137:129–136.
60. Kalmodia S, Molla AR, Basu B. *In vitro* cellular adhesion and antimicrobial property of $SiO_2$–MgO–$Al_2O_3$–$K_2O$–$B_2O_3$–F glass ceramic. *J Mater Sci Mater Med* 2010;21:1297–1309.
61. Cabal B, Alou L, Cafini F, Couceiro R, Sevillano D, Esteban-Tejeda L, Guitián F, Torrecillas R, Moya JS. A new biocompatible and antibacterial phosphate free glass-ceramic for medical applications. *Sci Rep* 2014;4:5440.
62. Chatzistavrou X, Fenno JC, Faulk D, Badylak S, Kasuga T, Boccaccini AR, Papagerakis P. Fabrication and characterization of bioactive and antibacterial composites for dental applications. *Acta Biomater* 2014;10: 3723–3732.

# Chapter 9

# Nanocomposites for Bone Repair

**Kyung-Ah Kwon*, Jennifer H. Shepherd†, David V. Shepherd‡, Ismael Moreno-Gomez§ and Serena M. Best¶**

*Department of Materials Science and Metallurgy, University of Cambridge, 27 Charles Babbage Road, Cambridge CB3 0FS, UK*

**kak31@cam.ac.uk*
†*jhr37@cam.ac.uk*
‡*dvs23@cam.ac.uk*
§*im380@cam.ac.uk*
¶*smb51@cam.ac.uk*

**Abstract**

The nanoceramics discussed within this book have a vast array of applications, but one of the most significant is as a reinforcing phase or filler within nanocomposites. Moving away from a focus on the nanoceramics themselves, this chapter has a particular emphasis on the common degradable polymer matrices and how they are combined with calcium phosphate ceramics. These polymer–bioceramic composites attempt to mimic the structure of bone, making them suitable for development as potential bone substitute materials. These may address clinical need in various medical fields such as orthopaedic, dental, and maxillofacial surgery. Methods of nanocomposite production are considered in some detail, and the mechanical, biological, and degradation properties of the nanocomposites are also briefly discussed. A short introduction is provided into the recent advances and likely future trends of the field. The chapter aims to demonstrate the key role nanocomposites can play in the field of bone repair.

## 9.1 Introduction

Bone is a remarkable natural nanocomposite, with a complex structural hierarchy organised through a self-assembly mode.[1] At the nanoscale, it consists of a fibrillar collagen structure with embedded mineral, and its hierarchy then demonstrates an irregular, yet optimised, arrangement and orientation of the components (Figure 9.1).

Not only does bone provide mechanical support and serve as a reservoir for minerals, but it also exhibits a unique capability for self-repair. Often successful reconstruction can be achieved only with restoration of alignment and stable fixation of the bone. Where this fails, techniques to replace, restore, and regenerate bone

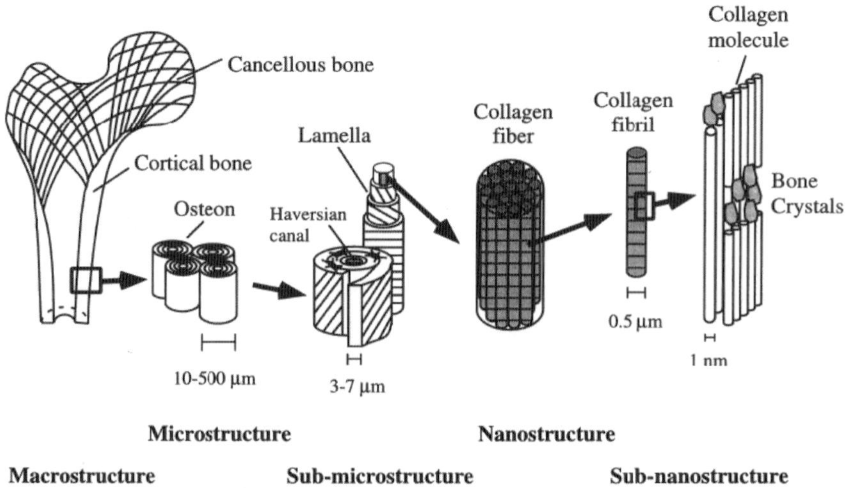


Figure 9.1 The hierarchical structure of bone. Reproduced with permission from Rho J-Y, Kuhn-Spearing L, Zioupos P. Mechanical properties and the hierarchical structure of bone. *Med Eng Phys* 1998;20:92–102. Copyright Elsevier.

have become a major clinical need in the fields of orthopaedic, spinal, dental, cranial, and maxillofacial surgery.

One of the most critical issues in tissue engineering is the realisation of scaffolds with specific physical, mechanical, and biological properties. Scaffolds must act as substrates for cellular growth and proliferation and provide support for new tissue formation.[2] At the simplest level, scaffolds should possess good biocompatibility, biodegradability with controllable degradation kinetics, easy fabrication, and sufficient mechanical properties.[2–5] Materials must exhibit both surface and structural compatibility.[3] In order to engineer functional tissues, scaffolds should be designed in order to facilitate cell distribution and guide tissue regeneration in three dimensions, degrading at a rate comparable to that of the new tissue formation.

Polymers are widely used in bone grafting as a result of their biocompatibility, design flexibility, availability of functional groups, ability for surface modification, low density, ease of fabrication, and ductility.[1] As Middleton and Tipton[4] summarise, an ideal polymer for any tissue engineering application would

- not evoke an inflammatory or toxic response, disproportionate to its beneficial effect;
- be metabolised without trace after fulfilling its purpose;
- be easily processed into its final form
- have an acceptable shelf life and be easily sterilised.

Polymers for tissue engineering can be broadly separated into synthetic and naturally derived polymers; both categories will be considered within this chapter. Because synthetic polymers can be produced under controlled conditions, they typically exhibit predictable and reproducible mechanical and physical properties such as mechanics and the degradation rate. It is also possible to control materials' impurities, reducing risks of immunogenicity, toxicity, and infection.[6] However, surfaces are often hydrophobic and lack cell-recognition signals.[2] Naturally derived polymers have the advantage of biological recognition which may support cell adhesion and function, but mechanics are generally more limited, supply is typically more restricted, and they tend to be expensive.[2]

Typically polymeric materials (particularly those exhibiting degradation) have insufficient stiffness and compressive strength compared to human bone — using either inorganic or organic fillers could be important to increase and modulate mechanical and degradation properties.[2] It is logical that two or more constituents offer different physical properties, which can be combined synergistically.[7] Properties and architecture of biological materials can be reflected more accurately by means of tailoring. Composite materials typically offer greater potential of structural biocompatibility than homogeneous so-called monolithic materials.[3] There are other reasons for development of polymer composite materials including: avoidance of corrosion and fatigue failure of metal alloys[8] and release of metal ions,[9] patient discomfort, and allergic reactions[10] and the low fracture toughness of ceramic implants. Biodegradable polymer materials typically degrade to acidic degradation products which although easily removed through biological processes, do nonetheless cause some degree of inflammatory reaction.[11] Addition of basic filling materials such as calcium phosphates may neutralise these acidic degradation products, increasing bioactivity.[12] Polymer composite materials can have their radio transparency adjusted by the addition of contrast media and are fully compatible with CT and MRI imaging.[3]

The first polymer matrix composites were neither biodegradable nor consisted of a nanoscale reinforcing phase. HAPEX™[13] consisted of a hydroxyapatite (HA)-reinforced polyethylene composite optimised as a bone analogue material; bone apposition was able to occur without fibrous encapsulation with the cell processes of human osteoblast cells observed to attach to the HA particles (Figure 9.2). Hydroxyapatite-reinforced polyurethane composites have also been considered.[14] Significant research has also been carried out on degradable polymer matrix composites with micronscale reinforcing ceramic phases[15,16] which are outside the scope of this chapter.

The transition from microparticles to nanoparticles can yield dramatic changes in the properties of the composite. Composites with nanoscale reinforcing particles (nanocomposites) have

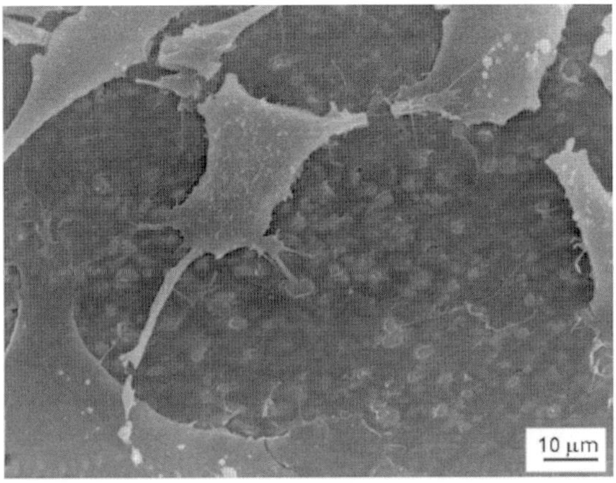


Figure 9.2 HOB cell processes attached to HA particles on the surface of HAPEX™. Reproduced with permission from Huang J, Di Silvio L, Wang M, *et al.* *In vitro* mechanical and biological assessment of hydroxyapatite-reinforced polyethylene composite. *J Mater Sci Mater Med* 1997;8:775–779. Copyright Springer.

generally been reported to exhibit enhanced properties in various aspects as compared with their counterparts containing micron-scale reinforcing particles (microcomposites). For example, two different composite films were prepared by Caridade *et al.*[17] by incorporating chitosan (CS) with 30 wt.% of bioactive glass (BG) — one with ~5 μm BG (μBG) and the other with 30–50 nm BG (nBG). When their mechanical properties were compared by performing dynamic mechanical analysis in tensile mode, the CS/nBG composite exhibited greater storage modulus than the CS/μBG composite. In addition, the CS/nBG composite was shown to have enhanced hydrophilicity measured by water contact angle and enhanced *in vitro* bioactivity assessed by examining apatite formation ability after immersing in simulated body fluid (SBF) for up to 7 days. The effect of using nanoscale filler as compared to micron-scale filler was also studied by others[18,19] who prepared three-dimensional porous composite scaffolds by adding either micron- or nanosized HA filler into synthetic polymers such as poly(ε-caprolactone) (PCL) and poly(L-lactic acid) (PLLA). Similar results of enhanced mechanical strength, increased hydrophilicity, and improved *in vitro* bioactivity assessed by the response of cells such as osteoblast-like cells, MG-63, or mesenchymal stem cells (MSCs) were reported. The incorporation of nanoscale filler was also shown to improve the surface topography, nanocomposite exhibiting improved nanotopography with more regular and homogeneous surface as compared with microcomposites.[20] Furthermore, the nanoscale reinforcement was shown to result in reduced acid release, indicating its higher buffering effect than micron-scale reinforcement.[21] These improvements can be attributed to relatively higher aspect ratio and specific surface area of the reinforcing particles in nanocomposites as compared to the ones in conventional microcomposites. The increase in surface area improves interactions between the polymer matrix and the nanostructure filler, resulting in the greater reinforcement effect and consequently enhanced performance in physical, mechanical, and biological properties.[2]

As the basic functional sub-units of cells and tissues are defined at the nanoscale, the application of nanotechnology represents a new frontier in tissue engineering research.[2] In order to translate understanding of nature into products useful in a clinical setting, chemical, physical, and biological properties need to be optimised in order to support, regulate, and influence long-term cellular activities.[5]

Whilst this chapter will consider primarily inorganic calcium containing fillers, it is important to note that a considerable body of research on nanocomposites considers both metallic and organic reinforcing phases. Metal nanocomposites have been typically used for visualisation, drug delivery, diagnosis, and therapy, whilst silver nanoparticles have been investigated for their antibacterial effect.[22] Carbon nanostructures are clearly the most celebrated products of nanotechnology to date, and nanotubes and graphene structures have been considered in various biocomposite applications.[23,24] Polymers may also of course form the fillers, for example, electrospun nanofibres as reinforcement in nanocomposite development.[25]

## 9.2 Nanocomposites

### 9.2.1 Polymer matrix

The decision of polymer matrix is dependent upon considerations of surface and structural compatibility. This means that not only should the implant surface be chemically, biologically, and physically suitable for the host tissue, but the mechanical behaviour should also be optimised for the host tissue.[1] This mechanical consideration is key not just upon implantation, but throughout the bone healing process. As bone healing progresses, it is desirable that bone is subjected to a gradual increase in stress,[3] additionally the desire must be to eliminate all synthetic material from the bone repair site. This can only occur through carefully controlled and understood degradation. Degradation kinetics are affected by: chemical composition and configurational structure, processing history, molar mass, polydispersity, environmental conditions, stress and strain, crystallinity, device size, morphology, chain orientation, distribution of chemically reactive compounds within the matrix, additives, presence of original monomers, and overall hydrophilicity.[6]

Biodegradable polymers are typically produced through the introduction of weak links within the polymer backbone — these are designed to permit controlled degradation of an initially high-molecular-weight hydrophobic polymer into lower-molecular-weight oligomer which can then be fully degraded *in vivo*.[6] Rezwan *et al.*[6] suggest some general guidelines which can be applied in the synthesis of biodegradable polymers:

- Higher ratio of hydrophilic/hydrophobic groups is better for degradation.
- Fully carbon chain polymers are less likely to degrade than heteroatom-containing polymers.
- Chain branching is deleterious to biodegradation.
- Lower-molecular-weight oligomers are more susceptible to biodegradation.
- Crystallinity reduces biodegradation.
- Lower surface area reduces biodegradation.

#### 9.2.1.1 *Saturated aliphatic polyesters (PLA, PGA, PCL, and their copolymers)*

The most commonly used biodegradable polymers are saturated poly-α-hydroxy esters; these include poly(lactic acid) (PLA), poly(glycolic acid) (PGA), and poly(lactic-co-glycolide) (PLGA) copolymers.[6,26] PCL is another example of saturated aliphatic polyester. The chemical properties of these polymers allow for hydrolytic degradation through de-esterification. Once degraded, the monomeric component of each polymer can be simply removed through natural pathways.[6] The body contains highly regulated mechanisms for removing monomeric components of glycolic acids (converted into glyoxylate and then glycine[27]) and lactic acid can of course be removed through the tricarboxylic acid cycle.

PLA and PGA can generally be processed easily and their degradation rates, physical, and mechanical properties are adjustable over a wide range by using various molecular weights and copolymers. They are commonly synthesised by a ring-opening polymerisation of the monomers.[26,28] However, these polymers can undergo an autocatalytic bulk erosion process which can cause scaffolds to fail prematurely and result in a strong inflammatory process.[11]

PGA, PLA, and their copolymers have been researched since the 1960s. Developing from degradable sutures,[4,29] uses have branched across dental,[30] orthopaedic,[31] and drug-delivery[32] applications.

PGA is the simplest linear aliphatic polyester, synthesised from the dimerisation of glycolic acid.[4] It is highly crystallised and typically insoluble in most organic solvents.[33] PGA typically exhibits high strength and modulus and is generally too stiff to be used as sutures except in a braided form.[4] PGA degrades rapidly in aqueous solution or *in vivo* and typically loses its mechanical integrity within 2–4 weeks.[28] Its degradation typically occurs in a two-stage process: the first involves diffusion of water into the amorphous regions of the matrix and hydrolytic chain scission of the ester groups, and the second stage then affects the crystalline regions.[26,34]

Although common processing techniques such as extrusion, injection, and compression moulding can be used to fabricate PGA into various forms, processing conditions require consideration as a result of its sensitivity to hydrolytic degradation. Solvent casting and particular leaching method can also be used to fabricate PGA-based implants.[26]

Poly(lactide) is the cyclic dimer of lactic acid and exists as two optical isomers D (DPLA) and L (LPLA).[35] Whilst the L form of PLA exhibits high tensile strength, low elongation, and high modulus due to its high crystallinity,[6] synthetic blends of D- and L-lactide are amorphous and hence have lower strength, higher elongation, and much more rapid degradation times. Whilst LPLA is used more commonly in orthopaedic fixation and sutures,[36,37] DLPLA is more attractive as a drug-delivery system.[4]

Through copolymerisation of PGA and PLA, it is possible to widely vary the degradation rate which is governed by the hydrophilic/hydrophobic balance and crystallinity of the respective polymer.[38] The highly amorphous nature of a PLGA copolymer (PGA and PLA in a 50/50 molar ratio) results in rapid degradation.[27,30]

PCL degrades much more slowly than other aliphatic polyesters and there appears no evidence of internal autocatalysis.[39] PCL is a semicrystalline polymer with a very low glass transition temperature (–60°C). It is therefore always in the rubbery state and shows high material permeability under physiological conditions.[28] Degradation mechanisms are typically similar to those of PLA — random hydrolytic ester cleavage and weight loss through diffusion of oligomeric species from the bulk.[6] However, degradation can also occur through enzymatic[40] or intracellular mechanisms.[41]

Table 9.1, modified from Armentano *et al.* summarises some of the key properties of these aliphatic polymers.

#### 9.2.1.2 *Polyhydroxyalkanoates* (*PHAs*)

Polyhydroxyalkanoates (PHAs) are aliphatic polyesters produced by microorganisms under unbalanced growth conditions.[6,42] Depending on the property requirements for the specific application, PHA polymers can either be blended, surface-modified, or copolymerised with other polymers, enzymes, or inorganic materials to further adjust mechanical properties or biocompatibility.[6]

Table 9.1 Key properties of the common biodegradable aliphatic polymer.

| | Thermal and mechanical properties | | | Degradation properties | | Processing and applications | |
|---|---|---|---|---|---|---|---|
| Polymer | Melting temp. (°C) | Glass transition temp. (°C) | Tensile modulus (GPa) | Time (month) | Deg. product | Solvent | Application |
| Polylactic acid (PLA) | 173–178 | 60–65 | 1.5–2.7 | 12–18 | L-lactic acid | Chloroform dioxane, dichloromethane, ethylacetate, acetone | Fracture fixation, interference screws, suture anchors, meniscus repair |
| Polyglycolic acid (PGA) | 225–230 | 35–40 | 5–7 | 3–4 | Glycolic acid | Hexafluoroisopropanol, acetone, dichloromethane, chloroform | Suture anchors, meniscus repair, medical devices, drug delivery |
| PCL | 58–63 | −60 | 0.4–0.6 | >24 | Caproic acid | Choloroform, Hexafluoroisopropanol, Dichloromethane, Toluene | Suture coating, dental, orthopaedic implants |
| Polylactic-co-glycolic PLGA (50/50) | Amorphous | 50–55 | 1.4–2.8 | 3–6 | D, L-lactic acid and glycolic acid | Choloroform, dichloromethane, ethylacetate, acetone, hexafluoroisopropanol | Sutures, drug delivery |

| | | | | | | | |
|---|---|---|---|---|---|---|---|
| Polylactic-co-glycolic PLGA (85/15) | Amorphous | 50–55 | 1.4–2.8 | 3–6 | D, L-lactic acid and glycolic acid | Dichlorometane Etylacetate Acetone Hexafluoroisopro panol | Interference screws, suture anchors, ACL reconstruction |
| Poly-lactic-co-glycolic PLGA (90/10) | Amorphous | 50–55 | — | <3 | D, L-lactic acid and glycolic acid | Choloroform, dichloromethane, ethylacetate, acetone, tetrahydrofuran | Artificial skin, wound healing, sutures |

Modified from Armentano I, Dottori M, Fortunati E, *et al*. Biodegradable polymer matrix nanocomposites for tissue engineering: A review. *Polym Degrad Stabil* 2010;95:2126–2146, with kind permission from Elsevier.

PHB (polyhydroxybutyrate) is of particular interest for bone tissue applications as it has been demonstrated to produce a consistent favourable bone tissue adaptation response with no evidence of an undesirable chronic inflammatory response after an implantation time of 12 months. The material also showed no evidence of extensive structural breakdown.[43]

#### 9.2.1.3 *Polyanhydrides*

The detrimental effects of bulk erosion processes in aliphatic polyesters have led to a drive to produce high-strength, surface-eroding polymers. Muggli *et al.*[44] considered cross-linked polyanhydrides for orthopaedic applications. They considered the polymerisation of methacrylated anhydride monomers of sebacic acid and 1,6-bis(carboxyphenoxy) hexane. The combination of hydrophobicity of the polymer backbone with hydrolytically anhydride linkages led to surface-eroding networks. By varying the amounts of the two monomers, degradation rates could be varied from 2 days to 1 year. Tensile moduli within the range of cortical and trabecular bone were also reported.[44] These materials are biocompatible and degrade into non-toxic diacid by-products that can be eliminated from the body as metabolites.[28]

#### 9.2.1.4 *Polyurethanes (PUs)*

Polyurethanes (PUs) remain one of the most popular groups of biomaterials and are used for a wide range of biomedical applications.[45] They are highly versatile in terms of mechanical and physical properties, biological properties, blood and tissue compatibility, and more recently degradability.[28] Whilst typically considered inert, biodegradable PUs can be synthesised through the incorporation of hydrolysable segments into their backbones.[45] Degradable polyurethane scaffolds have been shown to support the ingrowth of cells and tissues both *in vitro* and *in vivo* and undergo controlled degradation to non-cytotoxic degradation products.[45]

#### 9.2.1.5 *Polymers for hydrogels*

Hydrogels are crosslinked hydrophilic polymers that are able to absorb large amounts of water without dissolution. Among the synthetic hydrogels, polyethylene glycol (PEG) is the most extensively studied for tissue engineering research.[28,46] However, PEG hydrogels are not degradable,[47] although commonly applied in copolymers with degradable materials.[48,49]

#### 9.2.1.6 *Copolymerisation*

The capabilities of synthetic polymers are vast with even more varied and optimised properties obtained through copolymerisation. With numerous copolymers considered both as biomaterials in their own right and as the matrices for nanocomposites, this chapter is not going to consider copolymerisation outside of discussion directly applied to nanocomposites.

#### 9.2.1.7 *Biodegradable natural polymers*

Whilst synthetic polymers have the advantages of supply and consistency, biopolymers are advantageous due to their similarities with extracellular matrices (ECMs), their specific degradation rates, and good biological performance.[50] Inspired by the nature of bone, three-dimensional (3D), structurally hierarchical constructs and nanocomposites that can comprise several levels of organisation, from the macroscopic tissue arrangement down to the molecular arrangement of the proteins are an ideal.[50]

Collagen along with its denatured form gelatine must be the preferred natural polymer matrix due to its close similarity to ECMs. They also include functional groups that can enhance osteoblast adhesion and migration.[51] Collagen is naturally degraded by metalloproteases, specifically collagenase, and serine proteases, meaning its degradation can be locally controlled by cells present in the engineered tissue.[52]

Other natural polymers considered both as monolithic single-phase materials and as matrix components of nanocomposites include: polysaccharides such as chitosan,[53,54] glycosaminoglycans, particularly hyaluronic acid,[55,56] and bacterial cellulose-based materials.[57]

### 9.2.2 Calcium phosphate-based fillers

Bio-nanoceramics are increasingly applied as the fillers in nanocomposites for bone repair. As a result of the significant detail in which these materials have previously been covered, this section will consider them again only briefly. There has been very little investigation of the bioinert alumina and zirconia or calcium sulphates as filler materials for nanocomposites.[58,59] Silica proves some promise as a reinforcing phase for nanocomposites,[60] but in this chapter, we will focus on calcium phosphates such as hydroxyapatite (HA), tricalcium phosphate (TCP) ($\alpha$ and $\beta$), and biphasic calcium phosphates (BCP) and bioglasses.

#### 9.2.2.1 *Nanoscale bioceramic fillers*

As mentioned often throughout this book, nanostructured materials typically offer much improved performance than larger particle sized counterparts. Nanoscale ceramics are able to offer significant ductility prior to failure[61] and have been shown to promote osteoblast adhesion and proliferation, osteointegration as well as the deposition of calcium-containing materials.[62] Characteristic features of the calcium phosphate or bioglass powders such as particle size, particle shape, and distribution are obviously keys to the overall properties of a nanocomposite. Although the typical production methods can result in ceramic particles on the nanoscale, more technical methods can also be applied to produce more specific morphologies and structures.

Synthetic calcium phosphates have been produced with specific and unusual morphologies using a range of water-soluble organic additives or surfactants.[63] For example, polyaspartic acid has been shown to absorb preferentially on specific faces of growing calcium phosphate particles resulting in significant crystal growth inhibition.[64] Nanofilaments have been produced using a variety of techniques including the use of double hydrophobic block copolymers as dispersed templates[65] or the use of reverse micelles.[63] Microemulsion techniques have also been used for the synthesis of nanoscale bioglass particles.[66] Other techniques for the production of specific nanoscale morphologies of bioglass[67] have included gas-phase synthesis[68] and laser spinning techniques.[69]

#### 9.2.2.2 *Hydroxyapatite*

Hydroxyapatite (HA) has the chemical formula $Ca_{10}(PO_4)_6(OH)_2$ and in 1926, De Jong proved through X-ray diffraction that the mineral component of bone was a form of HA.[70] The first reported use of HA for bone repair was in the 1950s.[71] One unit cell contains 44 atoms in a hexagonal structure and it fits into the $P6_3/m$ space group.

Bone mineral, being the ion store for the body,[72] contains many other ions. To mimic bone mineral more closely, it is thus essential to try and substitute other ions into HA. In the case of cationic substitutions, ions replace some of the calcium ions and with anionic substitutions the phosphate or hydroxyl groups are replaced. The effect of the substituted ions change the physical attributes of the as-produced material, but also enhances the biological properties.[73]

#### 9.2.2.3 *Tricalcium phosphates*

Tricalcium phosphate, TCP ($Ca_3(PO_4)_2$) has two general forms ($\alpha$ and $\beta$). Whilst $\alpha$-TCP has monoclinic symmetry and is in the $P2_1/a$ space group, $\beta$-TCP has rhombohedral symmetry and is in the $R3c$ space group. $\beta$-TCP is formed when precipitated powders (produced using similar routes to HA) are heat treated to >700°C, whilst $\alpha$-TCP form is stable at temperatures in excess of $\approx$1125°C.

Once $\alpha$-TCP is present, the powder needs to be quenched to prevent transformation to the $\beta$-TCP form.[74]

The two types of TCP vary in their degradation and bioactivity. Generally, $\alpha$-TCP is observed to demonstrate much more rapid dissolution than the lower-temperature $\beta$-phase.[75,76] The rapid dissolution of $\alpha$-TCP has been reported to result in reduced bone formation.[76] Generally, TCP is reported to have similar mechanisms to HA,[77,78] the $\beta$-form has been reported to exhibit a higher compressive strength than the $\alpha$-form as a result of micro-cracking which occurs during the high-temperature transformation.[79]

Due to the high-temperature transformations necessary for the synthesis of TCPs, TCPs with grain sizes less than submicron are not generally possible even with advanced sintering techniques such as spark plasma sintering and other assisted sintering techniques.[80] It is therefore essential to reduce the size of the particles using techniques such as milling.

#### 9.2.2.4 *Biphasic calcium phosphates*

These are a mixture of $\beta$-TCP and HA and were first investigated in the 1980s. Pioneered by Daculsi *et al.*[81] considerable research has continued to this day.[82–84] These bioceramics are typically produced by the heat treatment of a single calcium-deficient apatite to temperatures above 700°C to form a biphasic mixture. This biphasic material has been shown to exhibit enhanced mechanics compared to single-phase hydroxyapatite, believed to be a result of surface compression due to the $\beta$–$\alpha$ transformation at elevated temperatures.[85] They are used in orthopaedic or dental applications alone in dense or porous form[82,83,86] as well as fillers in nanocomposites.[87,88]

#### 9.2.2.5 *Bioglass*

Bioglass was designed as a result of a casual conversation on a bus. Hench was asked by an army colonel whether there were any materials that could survive the harsh environment of the human body. Research led to the development of a glass containing $SiO_2$, $Na_2O$, CaO, and $P_2O_5$ in specific proportions.[89] Bioglass had at the time unique properties in that it could bond directly to bone; existing material implants at the time were designed to be bioinert.[89]

There has been much research attempting to optimise the chemistry of Bioglass but the original form, 45S5, still remains one of the most effective. This is composed of 46.1 mol.% $SiO_2$, 24.4 mol.% $Na_2O$, 26.9 mol.% CaO, and 2.6 mol.% $P_2O_5$. The bonding of new bone to this material was found to be so effective in early trials that attempted removal of the material resulted in bone fracture.[90]

Bioglass' bioactivity stems from its solubility. Through a rapid exchange of Na with $H_3O$ from solution, the pH of the solution increases. This leads to an attack on the $SiO_2$ glass and the dissolution of silica into the solution and the formation of Si–OH on the glass surface. Through continued dissolution of the glass and the migration of the calcium and phosphate ions to the surface, an amorphous calcium phosphate layer forms; this continues to grow with the increased dissolution of the glass.[91]

Bioglass does have its disadvantages: it typically degrades much faster than bone develops and generally has poor mechanical properties.[89] It has, however, had significant clinical applications, particularly, in non-load-bearing applications. It has been used to replace the bones in the middle ear allowing for recovery of hearing[92] and has also been used to provide a stable ridge for dentures,[93] as a jaw filler in the form of PerioGlas®, as a bone filler for non-load-bearing sites (NovaBone®), and in toothpastes through NovaMin®.

Bioglass has great potential and flexibility in processing routes.[89] It can therefore be easily placed inside a polymer matrix to form a composite.[6,94]

## 9.3 Preparation of Nanocomposites

There exist various forms of polymer–bioceramic nanocomposites, depending on the end-use applications in the biomedical field. This section will outline some synthesis methods of producing such nanocomposites reported in the literature. Some ways of mixing the two different phases will be highlighted followed by how these mixtures can be used to fabricate different forms of nanocomposites.

### 9.3.1 Polymer–bioceramic mixing

Organic polymers and nanosized bioceramic particles can be mixed together via two main approaches: *ex situ* and *in situ* (Table 9.2). This mixture is further processed to produce various forms of nanocomposites as described in the following sections (Figure 9.3).

The *ex situ* approach is the most common and straightforward method to prepare polymer–bioceramic mixture since it merely involves a direct mixing of the two readymade component materials. The *ex situ* approach can be further divided into two processing methods, namely solution processing and melt processing, depending on the condition of a polymer at the point when the bioceramic particles are introduced. In solution processing, the particles are introduced into the solution, which is formed by dissolving the host polymer in a suitable solvent such as acetone, acetic acid, and distilled water. This mixture then undergoes a rigorous mixing by magnetic stirring,[95,96] torque rheometer,[97,98] ball milling,[99] or ultrasonication[100–102] to promote uniform dispersion of the filler particles within the polymer matrix. Instead of directly introducing the filler into the polymer solution, the filler particles may also be dispersed in the common solvent as for the polymer then get ultrasonicated, which might break up some agglomerated particles.[95,96] This ultrasonicated filler solution is then mixed with the polymer solution. It is noteworthy that ultrasonication can result in an overall decrease in the molecular weight of the polymer and change in the distribution.[103,104] Therefore, the ultrasonication process to improve the filler distribution needs to be performed carefully after selecting appropriate sonication power, temperature, and time.

Unlike solution processing, melt processing is free from any solvent since bioceramic particles are mixed into a polymer matrix in its molten state. This melt processing is often performed using

Table 9.2 A summary of methods for polymer–bioceramic mixing.

| Approach | *Ex Situ* | | *In Situ* | |
|---|---|---|---|---|
| Nature | Physical | | Chemical | |
| | **Solution processing** | **Melt processing** | | |
| **Processing method** | Bioceramic particles, whether as they are or after dispersing in the common solvent, are added into a polymer solution under rigorous mixing | Bioceramic nanoparticles are added into a melted polymer | Polymer or bioceramic particles are synthesised in the presence of the other | Both polymer and bioceramic nanoparticles are co-synthesised |
| **Advantage** | — Simple processability<br>— No limitation on the nature of host components | — Solvent-free<br>— Suitable for mass production | More precise control over the size and morphology of the product | |
| **Disadvantage** | Solvent residue | Somewhat limited to thermoplastic polymers | Solvent residue | |

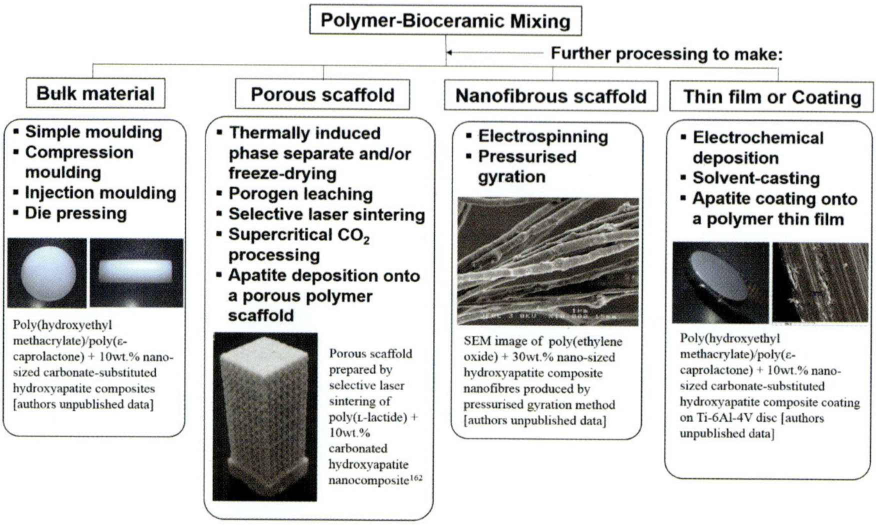


Figure 9.3 Some techniques to produce various forms of polymer–bioceramic nanocomposites.

an extrusion machine. The two components of nanocomposite, usually in powder or granule forms, are both fed into the hopper of the machine and then the polymer granules get melted in the barrel. Molten polymer and bioceramic particles are forced towards a rotating screw and extruded through a die forming a length of highly mixed product of constant cross-sectional area. Depending on the geometry of the die, various shapes such as "I-section", "O-section", and "square-section" can be produced. This continuous long end-product is cooled at the exit of the die by air or water and then gets cut into pellet forms for subsequent processes. Depending on the polymer properties such as degradation temperature, melting temperature, and molecular weight, the operating parameters such as screw configuration (single or twin), screw speed, residence time in the barrel, pressures and temperatures of feeding, melting, and mixing need to be optimised in order to obtain a high-performance end-product with homogeneous filler distribution. There are many papers in the literature, reporting the use of an extruder for blending and compounding of the polymer and bioceramic particles. For example, mixing polyethylene with 10–45 vol.% of hydroxyapatite,[105] mixing high-density polyethylene with 10–50 vol.% of glass-ceramic apatite–wollastonite (A–W)[106] and mixing ultra-high-molecular-weight polyethylene (UHMWPE) with 50 vol.% of hydroxyapatite.[107] However, a twin-screw extruder appears to be favoured over a single-screw extruder, probably due to the co-rotating or counter-rotating motion of the twin-screw being able to break down some aggregation of the bioceramic particles, promoting better dispersion of filler within the polymer matrix. Not only is this melt processing environmentally friendly, but it is also suitable for mass production at the industrial level.

While the *ex situ* approach involves using the preformed polymer and bioceramic particles to prepare a nanocomposite, the *in situ* approach involves chemical reactions to synthesise one or both of the nanocomposite components, i.e., one component is synthesised in the presence of the other or both components are co-synthesised. The former method can be further divided into two methods (*in situ* polymerisation or *in situ* inorganic particle synthesis), depending on which component of the nanocomposite is synthesised during the processing.

For *in situ* polymerisation, the monomer solution is first mixed with bioceramic nanoparticles, and it is then polymerised by free radicals or radiation, leading to nanocomposite production. For example, Li *et al.*[100] prepared a biodegradable nanocomposite of poly(methacrylated pyromellitylimidoalanine) (PM(PMA-ala)) and HA by first producing nanosized HA particles and then dispersing them into the anhydride monomer, M(PMA-ala) by ultrasonication. In the presence of the HA particles, the monomer gets polymerised via a photo-polymerisation process initiated by ultraviolet light with the photoinitiator, 2-dimethoxy-2-pheylacetophenone (DMPA). The similar approach has been reported by Huang *et al.*[108] who prepared a nanocomposite, poly-2-hydroxyethylmethacrylate)/$\varepsilon$-polycaprolactone (PHEMA/PCL) with HA. After melting PCL in the monomer, HEMA, 20 or 40 wt.% of nanosized HA was added, and subsequently the monomer gets polymerised via a free-radical polymerisation process with benzoyl peroxide as the initiator and ethylene glycol dimethacrylate as a crosslinking agent.

For *in situ* inorganic particle synthesis, bioceramic nanoparticles are synthesised in the presence of a polymer through chemical reactions, most typically via a sol–gel process, where inorganic precursors undergo *in situ* hydrolysis and condensation in the polymer solution or polymer melt. As a result, the inorganic precursors nucleate and grow, forming bioceramic particles in the presence of the polymer and consequently synthesising a nanocomposite. Murugan and Ramakrishna[109] made chitosan–HA nanocomposite by dispersing the HA precipitate into a 5 or 10 wt.% chitosan solution and allowing the hydrolysis-condensation reactions to continue to form a nanocomposite. The authors used calcium chloride ($CaCl_2$) and ammonium hydrogen phosphate $(NH_4)_2HPO_4$ as the HA precursors. A similar method has been reported by other authors as a way to synthesise chitosan–HA nanocomposite although different precursors such as $Ca(NO_3)_2.4H_2O$, $KH_2PO_4$, and $CaHPO_4.2H_2O$, were used for the HA production.[53] By varying the inorganic precursor and the polymer, a range of nanocomposites can be synthesised.[110–112] Martín *et al.*[111], for example, synthesised poly(vinyl alcohol) (PVA)-bioactive glass ($CaO–P_2O_5–SiO_2$) nanocomposite by adding PVA solution into the bioactive glass precursor solution, containing tetraethyl orthosilicate, triethyl phosphate, and calcium nitrate tetrahydrate for $SiO_2$, $P_2O_5$, and CaO, respectively. Furthermore, a nanocomposite can be produced by *in situ* co-precipitation method, in which a polymer solution containing phosphate acid ($H_3PO_4$) is added dropwise into calcium hydroxide ($Ca(OH)_2$) suspension under vigorous stirring. A nanocomposite can be obtained by ageing the precipitate and filtering followed by further processing to form a desired shape.[113,114] Instead of adding the inorganic precursors into a polymer solution, which is prepared by dissolving the polymer in a suitable solvent, the precursors can be added into a molten polymer to synthesise a nanocomposite. Dou *et al.*[115] used this approach to synthesise polypropylene (PP)–silica ($SiO_2$) nanocomposite using a co-rotating twin-screw microcompounder, where the silica being synthesised through the *in situ* hydrolysis and condensation of the precursor, polyethoxysiloxane by water vapour supplied with nitrogen flow.

Alternatively, a polymer–bioceramic mixture can be prepared *in situ* by simultaneously producing a polymer and bioceramic particles through synchronous monomer polymerisation and sol–gel reactions of inorganic precursors. He *et al.*[116] produced PEG–silica ($SiO_2$) composite using this method. Similarly, Kalkan-Sevinc and Goettler[117] used this route to produce nylon 66 (PA66)–$SiO_2$ nanocomposite.

During the *ex situ* approach, the physical mixing usually results in the mechanical locking of bioceramic particles within the polymer matrix,[107] leading to a weak interfacial bonding between the organic and inorganic phases through hydrogen bond and van der Waals interactions. However, this interfacial strength can be enhanced by surface grafting the filler with organic polymers such as PLLA and PCL.[118,119] This modification has been reported to improve both mechanical and biological properties of the composite. Similarly, a coupling agent such as 3-isocyanatopropyl triethoxysilane can be used during the sol–gel process to make more intimately mixed PCL–silica composites.[120]

### 9.3.2 Bulk material production

Bulk materials are typically made via a moulding technique since a desired shape, regardless of its complexity, can be easily manufactured with a high precision. Three types of moulding processes are described here, which are simple moulding, compression moulding, and injection moulding.

Simple moulding is usually performed by a liquid form of polymer–bioceramic mixture and it involves casting the mixture into a mould (glass, Teflon, silicones, polystyrene, etc.) and subsequent drying until the material solidifies.[53,108] In compression moulding, the process starts with heating of a mould to the desired temperature, and a slurry or powder form of the composite mixture is then introduced into the heated mould. A bulk composite material is formed by closing the mould under high pressure, compressing the mixture into the shape of the mould.[101,102,107] While the polymer–bioceramic mixture is directly introduced into a mould in compression moulding, the mixture is introduced into a barrel in injection moulding. This barrel heats up to the desired temperature, thereby melting the polymer–bioceramic mixture and then the molten material is delivered into a mould through the barrel nozzle by a high injection pressure. A high-performance nanocomposite can be obtained by optimising the processing parameters such as pressure and time of injection and holding (follow-up) process, plunger or screw speed and barrel and mould temperatures.[97,98] In addition to this conventional injection moulding, some authors reported using shear controlled orientation in injection moulding (SCORIM) to make a composite with anisotropic characteristic.[121] SCORIM equips with two oscillating pistons which continuously displace the melt mixture in the mould, resulting in the generation of a shear stress field at the melt/solid interface during holding period and consequently the anisotropic characteristic is induced in the end-product. Enhanced unidirectional mechanical properties were reported by the authors. The anisotropic characteristic can also be achieved by hot drawing process after the moulding process[107] since the post drawing process allows the polymer chains in the composite to be oriented along the drawing direction.

In addition to moulding technique, uniaxial or die pressing is widely used to make a bulk composite material for a powder form of the polymer–bioceramic mixture. It is a simple and cost-effective method but it is limited to manufacture a simple geometry.[101,114] This method involves compacting the powder into a bulk material by applying predetermined uniaxial mechanical or hydraulic pressure through a rigid punch.

### 9.3.3 Porous scaffold production

Porous scaffold can be fabricated by several techniques, some of which are described here. The degree of porosity and the pore structures such as pore's interconnectivity, size, and distribution can be varied greatly, depending on the technique as well as the processing conditions and parameters.

Firstly, a porous scaffold can be fabricated by thermally induced phase separation and/or freeze-drying technique(s). The former method involves placing the liquid or slurry polymer–bioceramic mixture prepared by *ex situ* solution processing or *in situ* approach into a freezer at –20°C. This induces the solid–liquid phase separation, i.e., the separation between the nanocomposite and the solvent. Upon the complete solvent solidification, the frozen solvent is removed by a dehydration process of freeze-drying via sublimation with controlled temperature and pressure.[95,96,122] A freeze-drying process alone can also be used to produce a porous scaffold as reported by Hong *et al.* who obtained poly(L-lactic) acid/bioactive glass ceramic (PLLA/BGC) nanocomposite by freeze-drying the PLLA/BGC/dioxane solution at –80°C for 1 week.[47]

Secondly, a porous scaffold can be fabricated by a porogen leaching technique, which involves mixing a polymer–bioceramic mixture solution with porogen such as sodium phosphate heptahydrate dibasic salt. This mixture is subsequently casted into a mould and then the solvent is dried. The dried bulk product is then immersed in deionised water for a period of time in order to completely remove the porogen. As a result, a final porous nanocomposite scaffold is formed, the pore spaces once being occupied by the porogen.[108]

Thirdly, the selective laser sintering (SLS) technique involves a laser and computer-aided design (CAD) modelling. The desired shape and structure of a scaffold is first drawn using a CAD software and then it is exported to a .STL file format so that it can be uploaded in the SLS computer system. Based on the 3D model, a laser beam is applied at the precise locations of the polymer–bioceramic mixture powder bed. By the heat of the laser beam, the powders then get fused together and when the unsintered powders are brushed off a 3D object with the desired shape is revealed.[123–125] A complex shape can easily be fabricated with this method since the laser beam is scanned layer by layer, being carefully controlled by the computer software. Depending on the laser power and the scanning speed, a range of porous structure can be produced.

The fourth method of making a porous scaffold involves a physical foaming agent, namely supercritical $CO_2$. This supercritical $CO_2$ processing starts from saturating polymer–bioceramic mixture with $CO_2$ above its critical temperature (31.1°C) and critical pressure (7.37 MPa or 73.7 bar) by diffusing $CO_2$ into the mixture. However, when the pressure is suddenly dropped to an atmospheric pressure, a phase separation occurs between the polymer–ceramic mixture and $CO_2$. As a result, $CO_2$ is rapidly dissipated and consequently gas bubbles within the polymer–bioceramic mixture are generated. These gas bubbles nucleate and grow within the plasticised polymer–bioceramic mixture since the rapid depressurisation also decreases the glass transition temperature of the polymer and consequently increases the polymer chain flexibility. However, when $CO_2$ is depleted, the glass transition temperature increases, thereby vitrifying the polymer–bioceramic nanocomposite with porous structure. Since the pores are developed by the escape of $CO_2$ from the plasticised molten polymer–bioceramic mixture, the pore structure and porosity can be tuned by varying the saturation temperature, saturation pressure, and saturation time.[126,127]

Finally, a porous scaffold of polymer–bioceramic nanocomposite can be fabricated by depositing apatite onto a porous polymer scaffold via various routes. For example, Zhang and Ma fabricated a PLLA/apatite porous scaffold by immersing a PLLA foam, which had been prepared by thermally induced phase separation technique, in SBF at 37°C for periods of time. The authors reported *in situ* formation of carbonated apatite on the PLLA porous scaffold.[122] Likewise, Roether *et al.*[128] prepared a porous PDLLA/bioglass scaffold via a slurry-dipping technique, which involves dipping a PDLLA foam into a slurry made by dispersing bioactive glass particles in distilled and deionised water and withdrawing the foam at a controlled speed after the predetermined time. Instead of immersing a porous polymer scaffold in SBF, a nanocomposite porous scaffold can be produced by centrifuging the polymer scaffold with polymer–bioceramic slurry several times. Xiao *et al.*[129] used this slurry centrifugal filling technique to fabricate a porous nanocomposite of polycaprolactone–chitosan/nano-hydroxyapatite–polyvinyl alcohol. Furthermore, a porous nanocomposite scaffold can be made by electrochemically depositing hydroxyapatite onto a polymer foam.[130]

### 9.3.4 Nanofibrous scaffold production

Polymeric nanofibres can be fabricated via several methods such as drawing, template synthesis, phase separation, self-assembly, and electrospinning. However, electrospinning appears to be the most popular choice in producing nanofibrous scaffolds for bone tissue engineering applications owing to its cost-effectiveness and capacity of generating long and continuous nanofibres with 3–1000 nm in fibre diameter and up to several metres in fibre length.[131]

In electrospinning, a polymer solution or polymer–bioceramic mixture solution is loaded in a plastic syringe, which is connected to a stainless steel needle through a capillary tube. Using a syringe pump, the solution is fed into the needle at a controlled rate and then a high DC voltage in the range of kV is applied to the needle, thereby inducing charges to the solution. At the tip of the needle, the solution is held by its surface tension as a hemispherical shape. However, as the voltage increases, this hemispherical shape changes into a conical shape known as the Taylor cone and the further increase in voltage results in the ejection of the solution as a jet from this Taylor cone to the grounded collector since the repulsive electrostatic force becomes stronger than the surface tension of the solution. While travelling towards the collector, the ejected jet gets elongated, evaporating the solvent, thereby forming long and thin nanofibres. The dimension and the morphology of nanofibres can be tailored by altering the solution parameters such as the viscosity, molecular weight of the polymer, filler content, and surface tension as well as processing conditions such as voltage, solution feeding rate, and distance between the tip of the needle and the grounded collector.[131] A number of researchers reported fabricating various polymer–bioceramic nanofibrous scaffolds using this method.[132–134]

More recently, the researchers at University College London (UCL) in the UK developed a new method for producing polymer nanofibres.[135] In this method called pressurised gyration process, they simply pour a polymer solution into an aluminium cylindrical vessel which has about 20 $\mu$m-sized orifices around its surface. While one end of the vessel is connected to a motor to allow it to rotate up to 36,000 rpm, the other end is connected to a nitrogen gas stream to provide a pressure up to 0.3 MPa. They reported that the fast rotating motion formed a polymer jet, which was subsequently stretched into fibres through the orifices. The fibres were observed to be stretched further when the gas stream was applied simultaneously. They reported producing nanofibres with 60–1000 nm in diameter with greater yield as compared with other nanofibre production techniques.[135] Considering the polymer solution needs to be simply poured into the vessel, this method could be easily applied for the polymer–bioceramic mixture solution to produce nanocomposite nanofibres.

### 9.3.5 *Thin film or coating production*

A thin film or coating of a polymer–bioceramic nanocomposite can be fabricated by several routes and electrophoretic deposition, also known as electrochemical deposition, is one among them. In this technique, the film or coating of a nanocomposite is formed by the deposition of charged particles onto an electrode under the influence of electric field. The coating thickness can be varied by altering the processing parameters such as the applied voltage and the deposition time. Moreover, this technique can be applied to a range of shapes from a plate to a complex 3D porous structure.[136] Pang and Zhitomirsky[137] produced chitosan/hydroxyapatite (CS/HA) coatings of thickness up to 50 $\mu$m by performing cathodic deposits on a range of substrates such as stainless steel wires and graphite plates from the suspension of CS/HA dispersed in ethanol–water mixture. Likewise, Wang *et al.*[138] fabricated chitosan/calcium phosphate (CS/

CaP) coatings on cathodic Ti6Al4V plates by performing the depositing process with a chitosan–supersaturated CaP mixture solution at 52°C for 15 h.

Solvent casting is a technique, which is commonly used to fabricate thin films due to its simplicity. A nanocomposite film can be simply made by casting a polymer–bioceramic mixture solution onto a Petri dish or a mould followed by the solvent evaporation. Upon complete evaporation of the solvent, the thin film can be retrieved since it peels off itself.[139] A nanocomposite film can also be fabricated by coating a polymer film with bioceramic. For example, Varma *et al.*[140] made a nanocomposite film by forming a porous hydroxyapatite coating on a chitosan film. The authors achieved this by first preparing a chitosan film through solvent casting and phosphorylating it using urea and phosphoric acid. The phosphorylated chitosan film was then soaked in saturated calcium hydroxide solution for 8 days followed by immersion in a concentrated SBF solution for more than 20 days.

## 9.4 Properties of Nanocomposites

The addition of nanostructured calcium-based ceramic particles to a polymeric matrix, either synthetic or natural, with the aim of improving its properties is a common approach in the biomedical field. There are a vast array of polymer matrix, calcium phosphate-reinforcing phase combinations, with Table 9.3 including a selection of those most widely researched.

For clarity and conciseness, this chapter focuses on biphasic nanocomposites made from a monophasic matrix and a monolithic filler. More complex nanocomposites, e.g., composites with three or more components, matrixes coated with nanocomposites, etc. have not been included.

Although the number of different combinations that can be found in literature is relatively high, there is a type that has received special attention: synthetic polymer–nanostructured hydroxyapatite composites. This system has proved to be one of the most popular, probably due to the extensive and wide use of synthetic

Table 9.3 Common combinations of polymer matrix and reinforcing calcium phosphate phase.

| | Filler | References |
|---|---|---|
| PLA | HA | 118, 134, 143–149 |
| | β-TCP | 150–154 |
| | α-TCP | 155 |
| | Bioglass | 156–159 |
| PLGA | HA | 133, 160–169 |
| | β-TCP | 170–173 |
| | α-TCP | 21, 174–178 |
| | Bioglass | 179, 180 |
| PCL | HA | 119, 181–185 |
| | β-TCP | 186–191 |
| | BG | 192 |
| Natural polymers (Collagen, gelatine, hyaluronic acid, chitosan, chitin) | HA | 53, 95, 193–198 |
| | β-TCP | 199–203 |
| | α-TCP | 204 |
| | Bioglass | 205–209 |
| PHB | HA | 210–212 |
| | Bioglass | 213–215 |

polymer on its own for medical applications[141] and the feasibility of obtaining hydroxyapatite nanoparticles using several methods.[142]

The following sections explore the effect of nanoparticulate ceramics in addition to a given matrix in terms of mechanical properties, biocompatibility, and degradation behaviour.

### 9.4.1 *Mechanical properties*

The addition of ceramic nanoparticles usually results in an improvement of some mechanical properties of the composites when compared with the polymeric matrix on its own.[216] In some cases, this change can be quite dramatic. Chang *et al.*[193] showed a 10-fold increase in Young's modulus for chitin hydrogels with the addition of nanohydroxyapatite (nHA) particles. In porous scaffolds for bone tissue engineering made of synthetic polymer, the addition of ceramic particles can increase the mechanical properties in compression mode up to four times.[145,152]

Deng *et al.*[144,181] reported a substantial improvement in the tensile modulus with the addition of nHA particles to the synthetic polymer matrix (PLA and PCL, respectively). The tensile modulus of the dense nanocomposites increased linearly with increasing filler content. In both cases, the yield stress showed little variation.[144,181] In the PCL composites, the fracture strain decreased with increasing filler content, suggesting a weak nanoparticulate–matrix interaction.[181] This weak interaction also results in a reduction of the bending strength as shown by Chen *et al.*[195] On the other hand, a good dispersion of the nanoparticles in the dense matrix improves the mechanical performance of the composites.[150]

In aligned electrospun nanofibrous PLGA/nHA scaffolds, dynamic mechanical analysis showed nHA to act as reinforcement at low concentrations (up to 5%) but as a defect at high concentrations (more than 10%). The storage modulus value of the scaffolds increased from 441 MPa for neat PLGA to 724 MPa for 5% nHA; however, further increasing the concentration led to a decrease in storage modulus, to 371 MPa for 20% nHA.[165]

Lee *et al.*[119] showed that the addition of HA nanoparticles increases the tensile strength of PCL composites although sometimes it causes a reduction in the ultimate elongation. This reduction can be avoided by grafting the ceramic particles. This surface modification of the nanoparticles results in a much greater interfacial bonding increasing the ultimate elongation.[119] On the same line, Cui *et al.*[160] showed an increase in compressive and bending strength in porous PLGA scaffolds with the addition of grafted nHA. This time, the increase was limited to a filler weight percentage of 20%. Higher percentages were not as effective.[160]

### 9.4.2 Bioactivity

In general, the addition of calcium-based ceramic nanoparticles to a given polymeric matrix increases the bioactivity. Hong *et al.*[118]

showed that the addition of nHA increases the biocompatibility by measuring the area occupied by adhered and proliferated chondrocytes. Moreover, this positive effect was shown to be intensified with the addition of grafted nHA nanoparticles.[118]

Lee examined the effect of increasing grafted nHA content into a PCL matrix in bioactivity using mouse NIH3T3 fibroblasts. The cell density after 7 days increased with increasing ceramic content.[119] Similarly, an increase in MG-63 proliferation with increasing $\beta$-TCP content was reported for porous PLLA/$\beta$-TCP nanocomposite scaffolds.[152] Additions of bioglass nanofiller greatly enhanced *in vitro* bone bioactivity by inducing calcium phosphate mineral on the fibre surface in electrospun PLA scaffolds for bone regeneration.[156] Osteoblastic cells (MC3T3-E1) adhered to and grew well on the nanocomposite membrane, with a higher level of growth than on the pure PCL membrane. The cells supported on the nanocomposite membrane expressed a significantly higher level of alkaline phosphatase (ALP) than those on the pure PCL membrane.[187]

### 9.4.3 Degradation behaviour

The degradation behaviour is a key property when considering nanocomposites for bone repair. The implants not only need to satisfy certain initial properties, but also need to maintain an appropriate behaviour over time. The overall degradation behaviour of a nanocomposite will be determined by the interplay between the individual responses of both the matrix and filler.

The degradation theory outlined here focuses on saturated aliphatic polyesters and its composites. The nanostructured filler will experience some degree of dissolution or dissociation when it is immersed in an aqueous medium. As a result, some negatively charged ions, as well as positively charged ions, will be released into the matrix.[217,218] For example, the TCP will dissociate into phosphate and calcium ions:

$$Ca_3(PO_4)_2 \rightarrow 3Ca^{2+} + 2PO_4^{3-}$$

The polymeric matrix will undergo chain cleavage activated by the aqueous medium and in some cases catalysed by the presence of hydrogen ions. For PLGA, the scission reaction can be expressed as

$$\text{PLGA} + H_2O \rightarrow \text{R-COOH} + \text{R-OH}$$

The carboxylic acid of the polyester has a high degree of dissociation, which results in the release of hydrogen ions and therefore an autocatalytic behaviour of the polymer degradation. The interaction of the two phenomena under certain circumstances can be expressed as the recombination of the phosphate ions in three consecutive steps in order to form phosphoric acid:

$$PO_4^{3-} + H^+ \rightarrow HPO_4^{2-}$$

$$HPO_4^{2-} + H^+ \rightarrow H_2PO_4^-$$

$$H_2PO_4^- + H^+ \rightarrow H_3PO_4$$

Therefore, it is envisioned that the addition of nanoceramic particles to the polymeric matrix in a dense composite will slow down the degradation rate. Yang *et al.*[21] showed that the addition of $\alpha$-TCP nanoparticles to a PLGA matrix significantly delays the onset of the pH decrease in the PBS buffer solution where the composites are immersed. This delay can be interpreted as the effect of a lower degradation rate meaning that the samples with nanoceramic filler later reached the critical molecular weight necessary for the release of the acidic internal degradation products into the media.[21]

Plenty of mathematical models have been developed to describe and ultimately predict the degradation of polymeric implants and the drug release rate for those loaded with drug.[217] These models use a number of techniques ranging from simple analytical solutions to 3D finite element modelling. Sadly, the complexity of the degradation phenomena in the case of nanocomposites means that modelling of composite degradation is a field yet to be fully explored.

Pan *et al.*[218] developed a model for the degradation of PLGA/α-TCP composites based on their previous model for poly-α-hydroxyacids degradation. In order to incorporate the effect of the ceramic filler, two sets of equations were incorporated into their previous chain scission model.

In this family of models for polymer degradation, the chain scission rate ($dR/dt$) can be expressed as

$$\frac{dR}{dt} = k_1 C_e + k_2 C_e C_{H^+}$$

where $C_e$, $C_{H^+}$, $k_1$, and $k_2$ are the ester bond concentration, the hydrogen ion concentration, the non-catalytic scission rate, and the catalytic scission rate, respectively.[219–221]

Amongst other parameters, the model outputs the change in molecular weight of the polymeric matrix and the variation of the amount of ceramic filler over time. Figure 9.4 depicts the evolution of the molecular weight of the polymeric matrix predicted by the model for both pure PLGA50:50 and PLGA/30 wt.% α-TCP composites with spherical particles of 4 μm and 50 nm in diameter. Figure 9.5 shows the experimental data used to generate these modelling results.

Although these models show promise, there is still room for improvement in order to increase the functionality of the models and incorporate aspects of the composite degradation such as swelling, ceramic phase change, polymer crystallisation, and ions transportation within the composite.

## 9.5 Journey to the Clinic

Despite the considerable body of research focused on the field of nanocomposites for bone repair, only Vitoss scaffold foam (Orthovita–Stryker), a combination of β-TCP and bovine type I collagen, and FortrOss (Pioneer Surgical), a combination of nanocrystalline HA and porcine collagen polymerised with dextran, have been applied clinically.[222] The absence of synthetic materials in these composites demonstrate the greater hesitance which still exists in the application of more alien materials to bone tissue engineering applications.

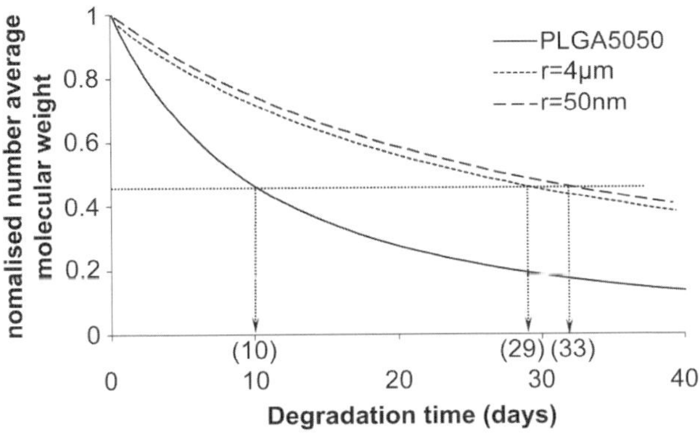


Figure 9.4 Predicted evolution of the normalised Mn (number-average molecular weight) for pure PLGA50:50 and PLGA/30 wt.% α-TCP composites with spherical particles of 4 μm and 50 nm in diameter. Reproduced with permission from Ref. 218. Copyright Elsevier.

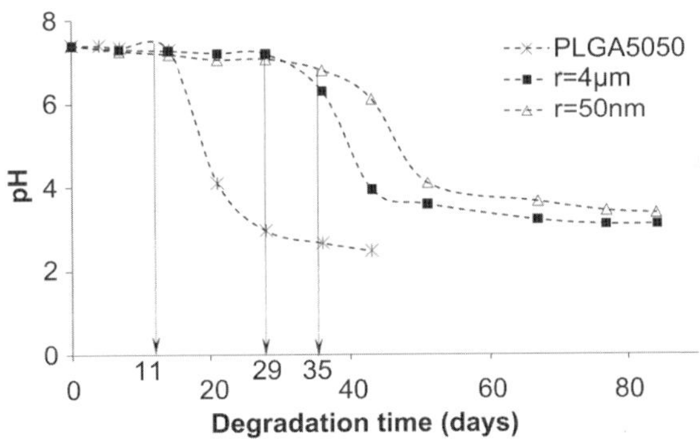


Figure 9.5 Change of the buffer media pH during degradation for pure PLGA50:50 and PLGA/30 wt.% α-TCP composites with spherical particles of 4 μm and 50 nm in diameter. The degradation time needed to reach the onset of the pH decrease is indicated for each material. Reproduced with permission from Ref. 218. Copyright Elsevier.

Vitoss in its path to clinical application has been implanted in a canine metaphyseal defect and the standard material was compared to the material also containing bioglass.[223] Both materials were shown to be biocompatible and support bone ingrowth; healing progression was time-dependent and the structure and stability of defect sites approached those of native cancellous bone after 24 weeks. However, more generally, *in vivo* models have considered small animals, typically rat and the New Zealand rabbit, and have often considered subcutaneous or muscle pouch implantation models.[166,224,225]

Published *in vivo* models on synthetic polymer matrix nanocomposites have generally demonstrated promise, although this may be attributed to the general tendency to publish only positive results. A study which implanted PCL/PLA/bioactive glass composite scaffolds in combination with bone marrow stromal cells (BMSCs) in a rat subcutaneous implantation model found that with and without bioglass, the scaffolds were fully invaded by well-vascularised soft connective tissue. Only with bioglass in combination with the BMSCs was significant mineral formation observed. Other studies with various synthetic polymer matrices and calcium phosphate fillers exhibited new bone formation after 2 months,[226] rapid and strong mineralisation,[169] good biocompatibility, and better performance in guided tissue regeneration than self-repair alone.[224] A hierarchically biomimetic bone scaffold material consisting of nano-HA, collagen, and PLA and designed to mimic bone in terms of both structure and chemistry, demonstrated bioactivity and after 12 weeks in a segmental defect in the rabbit radius, the implanted composite had been partially substituted by new bone tissue.[147]

There remains a significant need for more research in order to understand the precise mechanisms of nanocomposite–tissue interactions and also to optimise the composition, structure, and properties of different polymer composites. This must be carried out in order to fully achieve the potential of nanocomposites in bone tissue regeneration.[227] There is also clearly a need for effort to be spent on the application of nanocomposites in clinical and surgical settings, and this includes a significant body of *in vivo* investigations, including large animal trials before the preliminary clinical trials.

## 9.6 Recent Advances and Future Trends in Nanocomposites

The target of producing nanocomposites for bone repair remains to, as closely as possible, mimic the complex architecture of native bone tissue whether this is through synthetic or natural materials.[227] It has been suggested that strategic biomimicry should be the key.[228] With this approach, over-engineering is avoided, ideal matrix characteristics defined from nature, and then reproduced when it comes to scaffold design. In this section, we attempt to summarise some of the recent advances and future trends in the field of nanocomposites.

### 9.6.1 Combination with BMSCs

It has been suggested that engineered nanocomposite scaffolds, made with biodegradable polymer matrices, with nanostructure phases in combination with osteogenic growth factors or stem-cell seedings will play a vital role in bone repair.[2,229] Baler *et al.*[230] prepared a porous nanocomposite scaffold by adding 40 wt.% $\beta$-TCP into poly(1,8-octanediol-co-citrate) (POC). When the authors loaded the scaffold with bone morphogenic protein-2 (BMP-2), they found significantly increased biological response as compared with nanocomposite scaffold alone through both *in vitro* and *in vivo* studies. The differentiation of BMSCs can be controlled by the mechanics of the matrix material, encouraging the formation of specific tissue types. Engineered synthetic polymer nanocomposites can allow for precise and systematic control over mechanical properties of the cell substrate and it can also provide quantitative information about the forces sensed and exerted by cells.[2]

### 9.6.2 Electrical properties

Bone is a piezoelectric material,[231] which means an endogenous electric field is created in response to strain and this electrical stimulus has been reported to control cell growth and differentiation.[2,232] Shahini *et al.*[232] considered the production of a composite consisting of a biocompatible conductive polymer with gelatine and nanoscale bioglass. Increasing the concentration of conductive polymer was observed to enhance cell viability, indicating either improved scaffold structure or boosted electrical signalling.[232]

### 9.6.3 Vascularisation in bone tissue engineering constructs

The degree of vascularisation at bone defect sites is considered to be one of the important factors that affect the quantity of new bone formation and its growth.[233] Hence, vascularisation of large bone grafts remains one of the major challenges of bone tissue engineering. Several methods have been proposed to enhance vascularisation. For example, porous nanocomposite scaffold design to promote vascular ingrowth,[234,235] incorporation of angiogenic growth factors or cells such as vascular endothelial growth factor-releasing adipose-derived stromal cells,[236] and *in vitro* pre-vascularisation of a scaffold with endothelial progenitor cells or with co-culturing of endothelial cells and osteogenic cells.[237,238] Success of any graft depends on achieving full graft integration across multiple hierarchical levels, both between individual graft components as well as between implanted constructs and surrounding host tissue.[239] Mercado-Pagán *et al.*[239] suggest that a promising strategy for engineering vascularised bone tissue constructs is the incorporation of cell-laden hydrogels and synthetic tissue vascular grafts in a porous osteoconductive rigid frame in a hierarchical manner. The hydrogel is able to deliver signals and encapsulate and support cells necessary for osteogenesis and microvascularisation, the rigid frame offering improved structural guidance and mechanical integrity.

### 9.6.4 Nanocomposite hydrogels

Designing the next generation of nanocomposite hydrogels will require close control over physical, chemical, and electrical properties, but also the need to integrate suitable biological cues within the network.[240] As Gaharwar *et al.*[240] summarise multi-component networks will be necessary to impart multiple functionalities and it will prove key to understand the interaction between polymeric chains and nanoparticles at different length scales. New fabrication technologies will also of course be necessary in order to recapitulate the cellular micro-environment of native tissues within nanocomposite hydrogels.[240]

### 9.6.5 Natural-based nanocomposites

Future development of natural-based nanocomposites should be focused on a clear understanding of the nanocomposite–tissue interactions as well as the optimisation of composition and hierarchical structure in order to ensure long-term life.[50] Fatigue limit is a particular concern with natural-based composites and additionally use of nanocomposites with therapeutic effect and drug delivery combined with undifferentiated or differentiated cells should also be investigated.

### 9.6.6 Shape memory function

A material with shape memory capabilities would allow for delivery in compact form via minimally invasive surgery, and then upon contact with body temperature it would be able to recover its final complex state *in vivo*. Liu *et al.*[241] attempted to develop shape memory materials further through combining shape memory function with controlled delivery of growth factor BMP2 for repair of a mandibular bone defect.

### 9.6.7 Computer-aided scaffolding techniques

Whilst considerable focus is placed on the materials and structures of nanocomposites, methods of production must also be developed. Computer-aided scaffolding techniques provide significant advantages in terms of consistency, reproducibility of designed scaffolds, and capability of precise control over architecture of 3D scaffolds.[242] Several 3D computer-aided scaffolding techniques have been documented and were used to produce 3D polymer–bioceramic composite scaffolds. For example, 3D printing,[243] fused

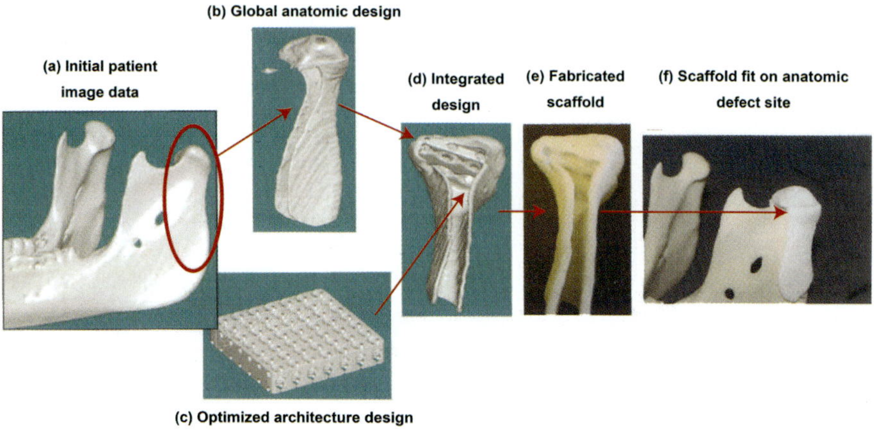


Figure 9.6 Image-based procedure for integrating designed microstructure with anatomic shape. Reproduced with permission from Ref. 248. Copyright Macmillan Publishers Ltd.

deposition modelling,[244] SLS,[123,145,245] and 3D plotting.[246] These techniques can be employed to produce a patient-specific scaffold based on the patient data such as CT or MRI scans in order to plan surgical procedures, to provide surgical templates or to offer personalised bone grafts/implants (Figure 9.6).[247,248] Further development will improve the resolution of these techniques, the breadth of the materials considered as well as the capabilities for printing biological factors and cells.

## 9.7 Summary

Bone is a highly vascular living tissue composed of an integrated structure of organic matrix and nanosized mineral components, thereby being identified as a natural nanocomposite. This chapter focused on nanocomposite materials, with a particular emphasis on those for bone repair. Some commonly used polymers, both synthetic and natural, are briefly described together with some calcium phosphate-based bioceramics, which mimic the organic and inorganic component of bone, respectively. Methods of mixing the two different components and subsequent nanocomposite production in various forms were highlighted in some detail. The mechanical, biological, and degradation properties of nanocomposite were discussed. Furthermore, special attention has been paid to the clinical application, recent advances, and future trends in nanocomposites.

## References


1. Murugan R, Ramakrishna S. Development of nanocomposites for bone grafting. *Compos Sci Technol* 2005;65(15–16):2385–2406.
2. Armentano I, Dottori M, Fortunati E, *et al.* Biodegradable polymer matrix nanocomposites for tissue engineering: A review. *Polym Degrad Stabil* 2010;95(11):2126–2146.
3. Ramakrishna S, Mayer J, Wintermantel E, *et al.* Biomedical applications of polymer–composite materials: A review. *Compos Sci Technol* 2001;61(9):1189–1224.
4. Middleton JC, Tipton AJ. Synthetic biodegradable polymers as orthopedic devices. *Biomaterials* 2000;21(23):2335–2346.
5. Akhilesh KG, Schexnailder PJ, Schmidt G. Nanocomposite polymer biomaterials for tissue repair of bone and cartilage: A material science perspective, in: Sitharaman B, editor. *Nanobiomaterials Handbook*. New York: CRC Press, Taylor & Francis Group; 2011; pp. 24-1–24-28.
6. Rezwan K, Chen QZ, Blaker JJ, *et al.* Biodegradable and bioactive porous polymer/inorganic composite scaffolds for bone tissue engineering. *Biomaterials* 2006;27(18):3413–3431.
7. Scholz M-S, Blanchfield JP, Bloom LD, *et al.* The use of composite materials in modern orthopaedic medicine and prosthetic devices: A review. *Compos Sci Technol* 2011;71(16):1791–1803.
8. Jacobs JJ, Gilbert JL, Urban RM. Current concepts review — Corrosion of metal orthopaedic implants. *J Bone Joint Surg Am* 1998; 80(2):268–282.
9. Savarino L. Release of metal ions in patients with implants. *J Bone Joint Surg Br* 2005;87-B(Suppl. I):55.
10. Hallab N, Merritt K, Jacobs JJ. Metal sensitivity in patients with orthopaedic implants. *J Bone Joint Surg Am* 2001;83-A(3):428–436.
11. Vert M, Li SM, Spenlehauer G, *et al.* Bioresorbability and biocompatibility of aliphatic polyesters. *J Mater Sci Mater Med* 1992;3:432–446.

12. Schiller C, Rasche C, Wehmöller M, *et al.* Geometrically structured implants for cranial reconstruction made of biodegradable polyesters and calcium phosphate/calcium carbonate. *Biomaterials* 2004;25(7–8):1239–1247.
13. Di Silvio L, Dalby MJ, Bonfield W. Osteoblast behaviour on HA/PE composite surfaces with different HA volumes. *Biomaterials* 2002;23(1):101–107.
14. Bos M, Van Dam GW, Jongsma T, *et al.* The effect of filler surface modification on the mechanical properties of hydroxyapatite-reinforced polyurethane composites. *Compos Interf* 1995;3(2):169–176.
15. Ehrenfried LM, Patel MH, Cameron RE. The effect of tri-calcium phosphate (TCP) addition on the degradation of polylactide-co-glycolide (PLGA). *J Mater Sci Mater Med* 2008;19(1):459–466.
16. Zhang R, Ma PX. Poly(a-hydroxyl acids)/hydroxyapatite porous composites for bone-tissue engineering. I. Preparation and morphology. *J Biomed Mater Res* 1999;44(4):446–455.
17. Caridade SG, Merino EG, Alves NM, *et al.* Chitosan membranes containing micro or nano-size bioactive glass particles: Evolution of biomineralization followed by *in situ* dynamic mechanical analysis. *J Mech Behav Biomed Mater* 2013;20:173–183.
18. Heo SJ, Kim SE, Wei J, *et al.* Fabrication and characterization of novel nano- and micro-HA/PCL composite scaffolds using a modified rapid prototyping process. *J Biomed Mater Res A* 2009;89(1):108–116.
19. Nejati E, Firouzdor V, Eslaminejad MB, *et al.* Needle-like nano hydroxyapatite/poly(l-lactide acid) composite scaffold for bone tissue engineering application. *Mater Sci Eng C* 2009;29(3):942–949.
20. Tamjid E, Bagheri R, Vossoughi M, *et al.* Effect of particle size on the *in vitro* bioactivity, hydrophilicity and mechanical properties of bioactive glass-reinforced polycaprolactone composites. *Mater Sci Eng C* 2011;31(7):1526–1533.
21. Yang Z, Best SM, Cameron RE. The influence of α-tricalcium phosphate nanoparticles and microparticles on the degradation of poly(D,L-lactide-co-glycolide). *Adv Mater* 2009;21(38–39):3900–3904.
22. Marambio-Jones C, Hoek EMV. A review of the antibacterial effects of silver nanomaterials and potential implications for human health and the environment. *J Nanoparticle Res* 2010;12(5):1531–1551.
23. Barrett JSF, Abdala AA, Srienc F. Poly(hydroxyalkanoate) elastomers and their graphene nanocomposites. *Macromolecules* 2014;47(12): 3926–3941.

24. Shi X, Hudson JL, Spicer PP, *et al.* Injectable nanocomposites of single-walled carbon nanotubes and biodegradable polymers for bone tissue engineering. *Biomacromolecules* 2006;7(7):2237–2242.
25. Huang Z-M, Zhang Y-Z, Kotaki M, *et al.* A review on polymer nanofibers by electrospinning and their applications in nanocomposites. *Compos Sci Technol* 2003;63(15):2223–2253.
26. Gunatillake PA, Adhikari R. Biodegradable synthetic polymers for tissue engineering. *Eur Cell Mater* 2003;5:1–16.
27. Hollinger JO. Preliminary report on the osteogenic potential of a biodegradable copolymer of polyactide (PLA) and polyglycolide (PGA). *J Biomed Mater Res* 1983;17(1):71–82.
28. Guo B, Ma PX. Synthetic biodegradable functional polymers for tissue engineering: A brief review. *Sci China Chem* 2014; 57(4): 490–500.
29. Auerbach R, Pearlstein MM. A comparison of polyglycolic acid (Dexon), nylon and silk sutures in skin surgery. *J Dermatol Surg Oncol* 1975;1(1 P2):38–40.
30. Miller RA, Brady JM, Cutright DE. Degradation rates of oral resorbable implants (polylactates and polyglycolates): Rate modification with changes in PLA/PGA copolymer ratios. *J Biomed Mater Res* 1977;11(5):711–719.
31. Vert M, Chabot F, Leray J, *et al.* Stereoregular bioresorbable polyesters for orthopaedic surgery. *Die Makromolekulare Chemie* 1981; 5(Suppl. 5):30–41.
32. Holland SJ, Tighe BJ, Gould PL. Polymers for biodegradable medical devices. 1. The potential of polyesters as controlled macromolecular release systems. *J Control Rel* 1986;4(3):155–180.
33. Bailey W, Schmitt E, Inventors; American Cyanamid Co, Assignee. Polyglycolic acid in solutions. Patent US3737440 A. Publication Date: 5th June 1973.
34. Chu CC. Hydrolytic degradation of polyglycolic acid: Tensile strength and crystallinity study. *J Appl Polym Sci* 1981;26(5): 1727–1734.
35. Okihara T, Tsuji M, Kawaguchi A, *et al.* Crystal structure of stereocomplex of poly(L-lactide) and poly(D-lactide). *J Macromol Sci B Phys* 1991;30(1–2):119–140.
36. Leenslag JW, Pennings AJ, Bos RRM, *et al.* Resorbable materials of poly(L-lactide). VI. Plates and screws for internal fracture fixation. *Biomaterials* 1987;8(1):70–73.

37. Mäkelä P, Pohjonen T, Törmälä P, *et al.* Strength retention properties of self-reinforced poly L-lactide (SR-PLLA) sutures compared with polyglyconate (Maxon) and polydioxanone (PDS) sutures. An *in vitro* study. *Biomaterials* 2002;23(12):2587–2592.
38. Reed AM, Gilding DK. Biodegradable polymers for use in surgery — poly(glycolic)/poly(Iactic acid) homo and copolymers: 2. *In vitro* degradation. *Polymer* 1981;22(4):494–498.
39. Lam CX, Hutmacher DW, Schantz JT, *et al.* Evaluation of polycaprolactone scaffold degradation for 6 months *in vitro* and *in vivo*. *J Biomed Mater Res A* 2009;90(3):906–919.
40. He F, Li S, Vert M, *et al.* Enzyme-catalyzed polymerization and degradation of copolymers prepared from ε-caprolactone and poly(ethylene glycol). *Polymer* 2003;44(18):5145–5151.
41. Woodward SC, Brewer PS, Moatamed F, *et al.* The intracellular degradation of poly(epsilon-caprolactone). *J Biomed Mater Res* 1985;19(4): 437–444.
42. Chen G-Q, Wu Q. The application of polyhydroxyalkanoates as tissue engineering materials. *Biomaterials* 2005;26(33):6565–6578.
43. Doyle C, Tanner ET, Bonfield W. *In vitro* and *in vivo* evaluation of polyhydroxybutyrate and of polyhydroxybutyrate reinforced with hydroxyapatite. *Biomaterials* 1991;12(9):841–847.
44. Muggli DS, Burkoth AK, Anseth KS. Crosslinked polyanhydrides for use in orthopedic applications: Degradation behavior and mechanics. *J Biomed Mater Res* 1999;46(2):271–278.
45. Guelcher SA. Biodegradable polyurethanes: Synthesis and applications in regenerative medicine. *Tissue Eng Part B Rev* 2008;14(1): 3–17.
46. Lin CC, Anseth KS. PEG hydrogels for the controlled release of biomolecules in regenerative medicine. *Pharm Res* 2009;26(3): 631–643.
47. Dobner S, Bezuidenhout D, Govender P, *et al.* A synthetic nondegradable polyethylene glycol hydrogel retards adverse postinfarct left ventricular remodeling. *J Card Fail* 2009;15(7):629–636.
48. Bryant SJ, Anseth KS. Controlling the spatial distribution of ECM components in degradable PEG hydrogels for tissue engineering cartilage. *J Biomed Mater Res A* 2003;64(1):70–79.
49. Saito N, Okada T, Horiuchi H, *et al.* A biodegradable polymer as a cytokine delivery system for inducing bone formation. *Nat Biotechnol* 2001;19(4):332–335.

50. Pina S, Oliveira JM, Reis RL. Natural-based nanocomposites for bone tissue engineering and regenerative medicine: A review. *Adv Mater* 2015;27(7):1143–1169.
51. Anselme K. Osteoblast adhesion on biomaterials. *Biomaterials* 2000;21(7):667–681.
52. Drury JL, Mooney DJ. Hydrogels for tissue engineering: Scaffold design variables and applications. *Biomaterials* 2003;24(24): 4337–4351.
53. Hu Q, Li B, Wang M, *et al.* Preparation and characterization of biodegradable chitosan/hydroxyapatite nanocomposite rods via *in situ* hybridization: A potential material as internal fixation of bone fracture. *Biomaterials* 2004;25(5):779–785.
54. Venkatesan J, Lowe B, Pallela R, *et al.* Chitosan-based polysaccharide biomaterials, in: Ramawat KG, Mérillon J-M, editors. *Polysaccharides.* Springer International Publishing, Switzerland; 2014; pp. 1–13.
55. Bakos D, Soldán M, Hernández-Fuentes I. Hydroxyapatite–collagen–hyaluronic acid composite. *Biomaterials* 1999;20(2):191–195.
56. Martínez-Sanz E, Ossipov DA, Hilborn J, *et al.* Bone reservoir: Injectable hyaluronic acid hydrogel for minimal invasive bone augmentation. *J Control Rel* 2011;152(2):232–240.
57. Qiu K, Netravali AN. A review of fabrication and applications of bacterial cellulose based nanocomposites. *Polym Rev* 2014;54(4): 598–626.
58. Han S-I, Lim JS, Kim DK, *et al. In situ* polymerized poly(butylene succinate)/silica nanocomposites: Physical properties and biodegradation. *Polym Degrad Stabil* 2008;93(5):889–895.
59. Hesaraki S, Moztarzadeh F, Nemati R, *et al.* Preparation and characterization of calcium sulfate–biomimetic apatite nanocomposites for controlled release of antibiotics. *J Biomed Mater Res B Appl Biomater* 2009;91(2):651–661.
60. Shchipunov YA, Karpenko TY. Hybrid polysaccharide–silica nanocomposites prepared by the sol–gel technique. *Langmuir* 2004;20(10): 3882–3887.
61. Kalita SJ, Bhardwaj A, Bhatt HA. Nanocrystalline calcium phosphate ceramics in biomedical engineering. *Mater Sci Eng C* 2007; 27(3):441–449.
62. Xu JL, Khor KA, Dong ZL, *et al.* Preparation and characterization of nano-sized hydroxyapatite powders produced in a radio frequency (rf) thermal plasma. *Mater Sci Eng A* 2004;374(1–2):101–108.

63. Sadasivan S, Khushalani D, Mann S. Synthesis of calcium phosphate nanofilaments in reverse micelles. *Chem Mater* 2005;17(10): 2765–2770.
64. Burke EM, Guo Y, Colon L, *et al.* Influence of polyaspartic acid and phosphophoryn on octacalcium phosphate growth kinetics. *Colloids Surf B* 2000;17(1):49–57.
65. Antonietti M, Breulmann M, Göltner CG, *et al.* Inorganic/organic mesostructures with complex architectures: Precipitation of calcium phosphate in the presence of double-hydrophilic block copolymers. *Chem Eur J* 1998;4(12):2493–2500.
66. Zhao NR, Wang YJ, Chen XF, *et al.* Preparation of bioactive nanoparticles in the system $CaO–P_2O_5–SiO_2$ using microemulsions. *Key Eng Mater* 2005;288–289:179–182.
67. Boccaccini AR, Erol M, Stark WJ, *et al.* Polymer/bioactive glass nanocomposites for biomedical applications: A review. *Compos Sci Technol* 2010;70(13):1764–1776.
68. Brunner TJ, Grass RN, Stark WJ. Glass and bioglass nanopowders by flame synthesis. *Chem Commun* 2006;(13):1384–1386.
69. Quintero F, Pou J, Comesaña R, *et al.* Laser spinning of bioactive glass nanofibers. *Adv Funct Mater* 2009;19(19):3084–3090.
70. de Jong WF. La substance minérale dans les os. *Recueil des Travaux Chimiques des Pays-Bas* 1926;45(6):445–448.
71. Ray RD, Degge J, Gloyd P, *et al.* Bone regeneration: An experimental study of bone-grafting materials. *J Bone Joint Surg Am* 1952; 34(3):638–647.
72. Mann S. Dalton perspectives. Biomineralization: The hard part of bioinorganic chemistry! *J Chem Soc Dalton Trans* 1993;(1):1–9.
73. Shepherd JH, Shepherd DV, Best SM. Substituted hydroxyapatites for bone repair. *J Mater Sci Mater Med* 2012;23(10):2335–2347.
74. Carrodeguas RG, De Aza S. $\alpha$-Tricalcium phosphate: Synthesis, properties and biomedical applications. *Acta Biomater* 2011;7(10): 3536–3546.
75. Nakagawa A, Matsuya S, Takeuchi A, *et al.* Comparison of the effects of added $\alpha$- and $\beta$- tricalcium phosphate on the basic properties of apatite cement. *Dent Mater J* 2007;26(3):342–347.
76. Yuan H, De Bruijn JD, Li Y, *et al.* Bone formation induced by calcium phosphate ceramics in soft tissue of dogs: A comparative study between porous $\alpha$-TCP and $\beta$-TCP. *J Mater Sci Mater Med* 2001; 12(1):7–13.

77. Akao M, Aoki H, Kato K. Mechanical properties of sintered hydroxyapatite for prosthetic applications. *J Mater Sci* 1981;16(3):809–812.
78. Akao M, Aoki H, Kato K, *et al.* Dense polycrystalline β-tricalcium phosphate for prosthetic applications. *J Mater Sci* 1982;17(2): 343–346.
79. Perera FH, Martínez-Vázquez FJ, Miranda P, *et al.* Clarifying the effect of sintering conditions on the microstructure and mechanical properties of β-tricalcium phosphate. *Ceram Int* 2010;36(6): 1929–1935.
80. Zhang F, Lin K, Chang J, *et al.* Spark plasma sintering of macroporous calcium phosphate scaffolds from nanocrystalline powders. *J Eur Ceram Soc* 2008;28(3):539–545.
81. Daculsi G, LeGeros RZ, Nery E, *et al.* Transformation of biphasic calcium phosphate ceramics *in vivo*: Ultrastructural and physicochemical characterization. *J Biomed Mater Res* 1989;23(8):883–894.
82. Daculsi G. Biphasic calcium phosphate concept applied to artificial bone, implant coating and injectable bone substitute. *Biomaterials* 1998;19(16):1473–1478.
83. LeGeros RZ, Lin S, Rohanizadeh R, *et al.* Biphasic calcium phosphate bioceramics: Preparation, properties and applications. *J Mater Sci Mater Med* 2003;14(3):201–209.
84. Ramay HR, Zhang M. Biphasic calcium phosphate nanocomposite porous scaffolds for load-bearing bone tissue engineering. *Biomaterials* 2004;25(21):5171–5180.
85. Royer A, Viguie JC, Heughebaert M, *et al.* Stoichiometry of hydroxyapatite: Influence on the flexural strength. *J Mater Sci Mater Med* 1993;4(1):76–82.
86. Nery EB, LeGeros RZ, Lynch KL, *et al.* Tissue response to biphasic calcium phosphate ceramic with different ratios of HA/β TCP in periodontal osseous defects. *J Periodontol* 1992;63(9):729–735.
87. Bakhtiari L, Rezaie HR, Hosseinalipour SM, *et al.* Investigation of biphasic calcium phosphate/gelatin nanocomposite scaffolds as a bone tissue engineering. *Ceram Int* 2010;36(8):2421–2426.
88. Nie L, Chen D, Suo J, *et al.* Physicochemical characterization and biocompatibility *in vitro* of biphasic calcium phosphate/polyvinyl alcohol scaffolds prepared by freeze-drying method for bone tissue engineering applications. *Colloids Surf B* 2012;100:169–176.
89. Jones JR. Review of bioactive glass: From Hench to hybrids. *Acta Biomater* 2013;9(1):4457–4486.

90. Hench LL, Splinter RJ, Allen WC, *et al.* Bonding mechanisms at the interface of ceramic prosthetic materials. *J Biomed Mater Res* 1971; 5(6):117–141.
91. Hench LL, Andersson Ö. Chapter 3: Bioactive glasses, in: Hench LL, Wilson J, editors. *An Introduction to Bioceramics*. Singapore: World Scientific; 1993; pp. 41–62.
92. Rust KR, Singleton GT, Wilson J, *et al.* Bioglass middle ear prosthesis: Long-term results. *Am J Otol* 1996;17(3):371–374.
93. Stanley HR, Hall MB, Clark AE, *et al.* Using 45S5 bioglass cones as endosseous ridge maintenance implants to prevent alveolar ridge resorption: A 5-year evaluation. *Int J Oral Maxillofac Implants* 1997;12(1):95–105.
94. Lu HH, El-Amin SF, Scott KD, *et al.* Three-dimensional, bioactive, biodegradable, polymer–bioactive glass composite scaffolds with improved mechanical properties support collagen synthesis and mineralization of human osteoblast-like cells *in vitro*. *J Biomed Mater Res Part A* 2003;64A(3):465–474.
95. Thein-Han WW, Misra RDK. Biomimetic chitosan–nanohydroxyapatite composite scaffolds for bone tissue engineering. *Acta Biomater* 2009;5(4):1182–1197.
96. Zhao F, Yin Y, Lu WW, *et al.* Preparation and histological evaluation of biomimetic three-dimensional hydroxyapatite/chitosan-gelatin network composite scaffolds. *Biomaterials* 2002;23(15): 3227–3234.
97. Cheang P, Khor KA. Effect of particulate morphology on the tensile behaviour of polymer–hydroxyapatite composites. *Mater Sci Eng A* 2003;345(1):47–54.
98. Bakar MSA, Cheang P, Khor KA. Mechanical properties of injection molded hydroxyapatite–polyetheretherketone biocomposites. *Compos Sci Technol* 2003;63(3–4):421–425.
99. Fang L, Leng Y, Gao P. Processing of hydroxyapatite reinforced ultrahigh molecular weight polyethylene for biomedical applications. *Biomaterials* 2005;26(17):3471–3478.
100. Li H, Chen Y, Xie Y. Photo-crosslinking polymerization to prepare polyanhydride/needle-like hydroxyapatite biodegradable nanocomposite for orthopedic application. *Mater Lett* 2003;57(19): 2848–2854.
101. Roeder RK, Sproul MM, Turner CH. Hydroxyapatite whiskers provide improved mechanical properties in reinforced polymer composites. *J Biomed Mater Res A* 2003;67(3):801–812.

102. Wang RZ, Cui FZ, Lu HB, *et al.* Synthesis of nanophase hydroxyapatite/collagen composite. *J Mater Sci Lett* 1995;14(7):490–492.
103. Tang ESK, Huang M, Lim LY. Ultrasonication of chitosan and chitosan nanoparticles. *Int J Pharmaceut* 2003;265(1–2):103–114.
104. Koda S, Mori H, Matsumoto K, *et al.* Ultrasonic degradation of water-soluble polymers. *Polymer* 1994;35(1):30–33.
105. Wang M, Joseph R, Bonfield, W. Hydroxyapatite–polyethylene composites for bone substitution: Effects of ceramic particle size and morphology. *Biomaterials* 1998;19(24):2357–2366.
106. Juhasz JA, Best SM, Brooks R, *et al.* Mechanical properties of glass-ceramic A–W-polyethylene composites: Effect of filler content and particle size. *Biomaterials* 2004;25(6):949–955.
107. Fang L, Gao P, Leng Y. High strength and bioactive hydroxyapatite nano-particles reinforced ultrahigh molecular weight polyethylene. *Composites Part B* 2007;38(3):345–351.
108. Huang J, Lin YW, Fu XW, *et al.* Development of nano-sized hydroxyapatite reinforced composites for tissue engineering scaffolds. *J Mater Sci Mater Med* 2007;18(11):2151–2157.
109. Murugan R, Ramakrishna S. Bioresorbable composite bone paste using polysaccharide based nano hydroxyapatite. *Biomaterials* 2004;25(17):3829–3835.
110. Davidenko N, Carrodeguas RG, Peniche C, *et al.* Chitosan/apatite composite beads prepared by *in situ* generation of apatite or Si-apatite nanocrystals. *Acta Biomater* 2010;6(2):466–476.
111. Martín AI, Salinas AJ, Vallet-Regí M. Bioactive and degradable organic–inorganic hybrids. *J Eur Ceram Soc* 2005;25(16):3533–3538.
112. Pereira MM, Jones JR, Orefice RL, *et al.* Preparation of bioactive glass–polyvinyl alcohol hybrid foams by the sol–gel method. *J Mater Sci Mater Med* 2005;16(11):1045–1050.
113. Chang MC, Ko CC, Douglas WH. Preparation of hydroxyapatite–gelatin nanocomposite. *Biomaterials* 2003;24(17):2853–2862.
114. Yamaguchi I, Tokuchi K, Fukuzaki H, *et al.* Preparation and microstructure analysis of chitosan/hydroxyapatite nanocomposites. *J Biomed Mater Res* 2001;55(1):20–27.
115. Dou Q, Zhu X, Peter K, *et al.* Preparation of polypropylene/silica composites by *in-situ* sol–gel processing using hyperbranched polyethoxysiloxane. *J Sol–Gel Sci Technol* 2008;48(1):51–60.
116. He J-P, Li H-M, Wang X-Y, *et al.* *In situ* preparation of poly(ethylene terephthalate)–$SiO_2$ nanocomposites. *Eur Polym J* 2006;42(5):1128–1134.

117. Kalkan-Sevinc ZS, Goettler LA. *In situ* polymerization of polyamide 66 nanocomposites utilizing interfacial polycondensation. III. Co-synthesis of silica nanocomposites via sol–gel chemistry. *Polym Eng Sci* 2012;52(11):2410–2416.
118. Hong Z, Zhang P, He C, *et al.* Nano-composite of poly(l-lactide) and surface grafted hydroxyapatite: Mechanical properties and biocompatibility. *Biomaterials* 2005;26(32):6296–6304.
119. Lee HJ, Kim SE, Choi HW, *et al.* The effect of surface-modified nano-hydroxyapatite on biocompatibility of poly(ε-caprolactone)/hydroxyapatite nanocomposites. *Eur Polym J* 2007;43(5):1602–1608.
120. Rhee SH, Choi JY, Kim HM. Preparation of a bioactive and degradable poly(ε-caprolactone)/silica hybrid through a sol–gel method. *Biomaterials* 2002;23(24):4915–4921.
121. Altpeter H, Bevis MJ, Gomes ME, *et al.* Shear controlled orientation in injection moulding of starch based blends intended for medical applications. *Plastics Rubber Compos* 2003;32(4):173–181.
122. Zhang R, Ma PX. Porous poly(L-lactic acid)/apatite composites created by biomimetic process. *J Biomed Mater Res* 1999;45(4):285–293.
123. Chua CK, Leong KF, Tan KH, *et al.* Development of tissue scaffolds using selective laser sintering of polyvinyl alcohol/hydroxyapatite biocomposite for craniofacial and joint defects. *J Mater Sci Mater Med* 2004;15(10):1113–1121.
124. Hao L, Savalani MM, Zhang Y, *et al.* Selective laser sintering of hydroxyapatite reinforced polyethylene composites for bioactive implants and tissue scaffold development. *Proc Inst Mech Eng H* 2006;220(4):521–531.
125. Zhou WY, Lee SH, Wang M, *et al.* Selective laser sintering of porous tissue engineering scaffolds from poly(L-lactide)/carbonated hydroxyapatite nanocomposite microspheres. *J Mater Sci Mater Med* 2008;19(7):2535–2540.
126. Mathieu LM, Montjovent MO, Bourban PE, *et al.* Bioresorbable composites prepared by supercritical fluid foaming. *J Biomed Mater Res A* 2005;75(1):89–97.
127. Mou Z-L, Zhao L-J, Zhang Q-A, *et al.* Preparation of porous PLGA/HA/collagen scaffolds with supercritical $CO_2$ and application in osteoblast cell culture. *J Supercrit Fluids* 2011;58(3):398–406.
128. Roether JA, Boccaccini AR, Hench LL, *et al.* Development and *in vitro* characterisation of novel bioresorbable and bioactive composite

materials based on polylactide foams and Bioglass® for tissue engineering applications. *Biomaterials* 2002;23(18):3871–3878.
129. Xiao X, Liu R, Huang Q. Preparation and characterization of nanohydroxyapatite/polymer composite scaffolds. *J Mater Sci Mater Med* 2008;19(11):3429–3435.
130. Huang Z-H, Dong Y-S, Chu C-L, *et al.* Electrochemistry assisted reacting deposition of hydroxyapatite in porous chitosan scaffolds. *Mater Lett* 2008;62(19):3376–3378.
131. Ramakrishna S, Fujihara K, Teo W-E, *et al. An Introduction to Electrospinning and Nanofibers*. Singapore: World Scientific; 2005.
132. Catledge SA, Clem WC, Shrikishen N, *et al.* An electrospun triphasic nanofibrous scaffold for bone tissue engineering. *Biomed Mater* 2007;2(2):142–150.
133. Lao L, Wang Y, Zhu Y, *et al.* Poly(lactide-co-glycolide)/hydroxyapatite nanofibrous scaffolds fabricated by electrospinning for bone tissue engineering. *J Mater Sci Mater Med* 2011;22(8):1873–1884.
134. Prabhakaran MP, Venugopal J, Ramakrishna S. Electrospun nanostructured scaffolds for bone tissue engineering. *Acta Biomater* 2009;5(8):2884–2893.
135. Mahalingam S, Edirisinghe M. Forming of polymer nanofibers by a pressurised gyration process. *Macromol Rapid Commun* 2013;34(14):1134–1139.
136. Boccaccini AR, Keim S, Ma R, *et al.* Electrophoretic deposition of biomaterials. *J R Soc Interf* 2010;7(Suppl 5):S581–S613.
137. Pang X, Zhitomirsky I. Electrodeposition of composite hydroxyapatite–chitosan films. *Mater Chem Phys* 2005;94(2–3):245–251.
138. Wang J, de Boer J, de Groot, K. Preparation and characterization of electrodeposited calcium phosphate/chitosan coating on Ti6Al4V plates. *J Dent Res* 2004;83(4):296–301.
139. Kumar R, Prakash KH, Cheang P, *et al.* Chitosan-mediated crystallization and assembly of hydroxyapatite nanoparticles into hybrid nanostructured films. *J R Soc Interf* 2008;5(21):427–439.
140. Varma HK, Yokogawa Y, Espinosa FF, *et al.* Porous calcium phosphate coating over phosphorylated chitosan film by a biomimetic method. *Biomaterials* 1999;20(9):879–884.
141. Cameron RE, Kamvari-Moghaddam A. Synthetic bioresorbable polymers, Chapter 3, in: Buchanan F, editor. *Degradation Rate of Bioresorbable Materials: Prediction and Evaluation*. Boca Raton, FL: Woodhead Publishing Limited and CRC Press LLC; 2008; pp. 43–66.

142. Ferraz MP, Monteiro FJ, Manuel CM. Hydroxyapatite nanoparticles: A review of preparation methodologies. *J Appl Biomater Biomech* 2004;2(2):74–80.
143. Aydin E, Planell JA, Hasirci V. Hydroxyapatite nanorod-reinforced biodegradable poly(L-lactic acid) composites for bone plate applications. *J Mater Sci Mater Med* 2011;22(11):2413–2427.
144. Deng X, Hao J, Wang C. Preparation and mechanical properties of nanocomposites of poly(D,L-lactide) with Ca-deficient hydroxyapatite nanocrystals. *Biomaterials* 2001;22(21):2867–2873.
145. Duan B, Wang M, Zhou WY, *et al.* Three-dimensional nanocomposite scaffolds fabricated via selective laser sintering for bone tissue engineering. *Acta Biomater* 2010;6(12):4495–4505.
146. Kim HW, Lee HH, Knowles JC. Electrospinning biomedical nanocomposite fibers of hydroxyapatite/poly(lactic acid) for bone regeneration. *J Biomed Mater Res A* 2006;79(3):643–649.
147. Liao SS, Cui FZ, Zhang W, *et al.* Hierarchically biomimetic bone scaffold materials: Nano-HA/collagen/PLA composite. *J Biomed Mater Res B Appl Biomater* 2004;69(2):158–165.
148. Spadaccio C, Rainer A, Trombetta M, *et al.* Poly-L-lactic acid/hydroxyapatite electrospun nanocomposites induce chondrogenic differentiation of human MSC. *Ann Biomed Eng* 2009;37(7):1376–1389.
149. Sui G, Yang X, Mei F, *et al.* Poly-L-lactic acid/hydroxyapatite hybrid membrane for bone tissue regeneration. *J Biomed Mater Res A* 2007;82(2):445–454.
150. Rakovsky A, Gotman I, Rabkin E, *et al.* Strong bioresorbable Ca phosphate–PLA nanocomposites with uniform phase distribution by attrition milling and high pressure consolidation. *J Mech Behav Biomed Mater* 2013;18:37–46.
151. Kankilic B, Bayramli E, Kilic E, *et al.* Vancomycin containing PLLA/$\beta$-TCP controls MRSA *in vitro*. *Clin Orthop Relat Res* 2011;469(11):3222–3228.
152. Lou T, Wang X, Song G, *et al.* Fabrication of PLLA/$\beta$-TCP nanocomposite scaffolds with hierarchical porosity for bone tissue engineering. *Int J Biol Macromol* 2014;69:464–470.
153. McCullen SD, Zhu Y, Bernacki SH, *et al.* Electrospun composite poly(L-lactic acid)/tricalcium phosphate scaffolds induce proliferation and osteogenic differentiation of human adipose-derived stem cells. *Biomed Mater* 2009;4(3):035002.
154. Makarov C, Berdicevsky I, Raz-Pasteur A, *et al.* *In vitro* antimicrobial activity of vancomycin-eluting bioresorbable $\beta$-TCP-polylactic acid

nanocomposite material for load-bearing bone repair. *J Mater Sci Mater Med* 2013;24(3):679–687.

155. Wilberforce SI, Finlayson CE, Best SM, *et al.* A comparative study of the thermal and dynamic mechanical behaviour of quenched and annealed bioresorbable poly-L-lactide/α-tricalcium phosphate nanocomposites. *Acta Biomater* 2011;7(5):2176–2184.
156. Noh K-T, Lee H-Y, Shin U-S, *et al.* Composite nanofiber of bioactive glass nanofiller incorporated poly(lactic acid) for bone regeneration. *Mater Lett* 2010;64(7):802–805.
157. El-Kady AM, Ali AF, Farag MM. Development, characterization, and *in vitro* bioactivity studies of sol–gel bioactive glass/poly(L-lactide) nanocomposite scaffolds. *Mater Sci Eng C* 2010;30(1):120–131.
158. Kim HW, Lee HH, Chun GS. Bioactivity and osteoblast responses of novel biomedical nanocomposites of bioactive glass nanofiber filled poly(lactic acid). *J Biomed Mater Res A* 2008;85(3):651–663.
159. Hong Z, Reis RL, Mano JF. Preparation and *in vitro* characterization of scaffolds of poly(L-lactic acid) containing bioactive glass ceramic nanoparticles. *Acta Biomater* 2008;4(5):1297–1306.
160. Cui Y, Liu Y, Cui Y, *et al.* The nanocomposite scaffold of poly(lactide-co-glycolide) and hydroxyapatite surface-grafted with L-lactic acid oligomer for bone repair. *Acta Biomater* 2009;5(7):2680–2692.
161. Aboudzadeh N, Imani M, Shokrgozar MA, *et al.* Fabrication and characterization of poly(D,L-lactide-co-glycolide)/hydroxyapatite nanocomposite scaffolds for bone tissue regeneration. *J Biomed Mater Res A* 2010;94(1):137–145.
162. Hong Z, Zhang P, Liu A, *et al.* Composites of poly(lactide-co-glycolide) and the surface modified carbonated hydroxyapatite nanoparticles. *J Biomed Mater Res A* 2007;81(3):515–522.
163. Leung LH, Fan S, Naguib HE. Fabrication of 3D electrospun structures from poly(lactide-co-glycolide acid)–nano-hydroxyapatite composites. *J Polym Sci Part B* 2012;50(4):242–249.
164. Lock J, Nguyen TY, Liu H. Nanophase hydroxyapatite and poly(lactide-co-glycolide) composites promote human mesenchymal stem cell adhesion and osteogenic differentiation *in vitro*. *J Mater Sci Mater Med* 2012;23(10):2543–2552.
165. Jose MV, Thomas V, Johnson KT, *et al.* Aligned PLGA/HA nanofibrous nanocomposite scaffolds for bone tissue engineering. *Acta Biomater* 2009;5(1):305–315.
166. Ngiam M, Liao S, Patil AJ, *et al.* The fabrication of nano-hydroxyapatite on PLGA and PLGA/collagen nanofibrous composite scaffolds

and their effects in osteoblastic behavior for bone tissue engineering. *Bone* 2009;45(1):4–16.
167. Naik A, Best SM, Cameron RE. The influence of silanisation on the mechanical and degradation behaviour of PLGA/HA composites. *Mater Sci Eng C* 2015;48:642–650.
168. Kim S-S, Park MS, Jeon O, *et al.* Poly(lactide-co-glycolide)/hydroxyapatite composite scaffolds for bone tissue engineering. *Biomaterials* 2006;27(8):1399–1409.
169. Zhang P, Hong Z, Yu T, *et al. In vivo* mineralization and osteogenesis of nanocomposite scaffold of poly(lactide-co-glycolide) and hydroxyapatite surface-grafted with poly(L-lactide). *Biomaterials* 2009;30(1):58–70.
170. Yu D, Li Q, Mu X, *et al.* Bone regeneration of critical calvarial defect in goat model by PLGA/TCP/rhBMP-2 scaffolds prepared by low-temperature rapid-prototyping technology. *Int J Oral Maxillofac Surg* 2008;37(10):929–934.
171. Yang YF, Tang GW, Zhao YH, *et al.* Effect of degradation of PLGA and PLGA/β-TCP scaffolds on the growth of osteoblasts. *Chin Sci Bull* 2011;56(10):982–986.
172. Kim H-M. Ceramic bioactivity and related biomimetic strategy. *Curr Opin Solid State Mater Sci* 2003;7(4–5):289–299.
173. Kulkova J, Moritz N, Suokas EO, *et al.* Osteointegration of PLGA implants with nanostructured or microsized β-TCP particles in a minipig model. *J Mech Behav Biomed Mater* 2014;40:190–200.
174. Barrett CE, Cameron RE. X-ray microtomographic analysis of α-tricalcium phosphate-poly(lactic-co-glycolic) acid nanocomposite degradation. *Polymer* 2014;55(16):4041–4049.
175. Wilberforce SI, Finlayson CE, Best SM, *et al.* The influence of the compounding process and testing conditions on the compressive mechanical properties of poly(D,L-lactide-co-glycolide)/α-tricalcium phosphate nanocomposites. *J Mech Behav Biomed Mater* 2011;4(7):1081–1089.
176. Sharma A, Meyer F, Hyvonen M, *et al.* Osteoinduction by combining bone morphogenetic protein (BMP)-2 with a bioactive novel nanocomposite. *Bone Joint Res* 2012;1(7):145–151.
177. Ege D, Cameron R, Best S. The degradation behavior of nanoscale HA/PLGA and α-TCP/PLGA composites. *Bioinspired Biomimetic Nanobiomater* 2014;3(2):85–93.
178. Barrett CE, Cameron RE. Nanoindentation analysis of α-tricalcium phosphate-poly(lactide-co-glycolide) nanocomposite degradation. *Mater Sci Eng C Mater Biol Appl* 2014;42:587–594.

179. Chen J, Fan X, Zhou Z, *et al.* The preparation and properties of bioactive composites based on modification bioactive glass and poly(lactide-co-glycolide). *Polym-Plastics Technol Eng* 2010;49(11): 1155–1162.
180. Hild N, Tawakoli PN, Halter JG, *et al.* pH-dependent antibacterial effects on oral microorganisms through pure PLGA implants and composites with nanosized bioactive glass. *Acta Biomater* 2013;9(11):9118–9125.
181. Hao J, Yuan M, Deng X. Biodegradable and biocompatible nanocomposites of poly(ε-caprolactone) with hydroxyapatite nanocrystals: Thermal and mechanical properties. *J Appl Polym Sci* 2002; 86(3):676–683.
182. Kim HW. Biomedical nanocomposites of hydroxyapatite/polycaprolactone obtained by surfactant mediation. *J Biomed Mater Res A* 2007;83(1):169–177.
183. Hao J, Liu Y, Zhou S, *et al.* Investigation of nanocomposites based on semi-interpenetrating network of [L-poly(ε-caprolactone)]/[net-poly(ε-caprolactone)] and hydroxyapatite nanocrystals. *Biomaterials* 2003;24(9):1531–1539.
184. Rezaei A, Mohammadi MR. *In vitro* study of hydroxyapatite/polycaprolactone (HA/PCL) nanocomposite synthesized by an *in situ* sol–gel process. *Mater Sci Eng C* 2013;33(1):390–396.
185. Lee HJ, Choi HW, Kim KJ, *et al.* Modification of hydroxyapatite nanosurfaces for enhanced colloidal stability and improved interfacial adhesion in nanocomposites. *Chem Mater* 2006;18(21): 5111–5118.
186. Ozkan S, Kalyon DM, Yu X. Functionally graded β-TCP/PCL nanocomposite scaffolds: *In vitro* evaluation with human fetal osteoblast cells for bone tissue engineering. *J Biomed Mater Res A* 2010; 92(3):1007–1018.
187. Erisken C, Kalyon DM, Wang H. Functionally graded electrospun polycaprolactone and β-tricalcium phosphate nanocomposites for tissue engineering applications. *Biomaterials* 2008;29(30): 4065–4073.
188. Bernstein M, Gotman I, Makarov C, *et al.* Low temperature fabrication of β-TCP–PCL nanocomposites for bone implants. *Adv Eng Mater* 2010;12(8):B341–B347.
189. Schneider OD, Weber F, Brunner TJ, *et al.* *In vivo* and *in vitro* evaluation of flexible, cottonwool-like nanocomposites as bone substitute material for complex defects. *Acta Biomater* 2009;5(5):1775–1784.

190. Damadzadeh B, Jabari H, Skrifvars M, *et al.* Effect of ceramic filler content on the mechanical and thermal behaviour of poly-L-lactic acid and poly-L-lactic-co-glycolic acid composites for medical applications. *J Mater Sci Mater Med* 2010;21(9):2523–2531.
191. Makarov C, Gotman I, Radin S, *et al.* Vancomycin release from bioresorbable calcium phosphate–polymer composites with high ceramic volume fractions. *J Mater Sci* 2010;45(23):6320–6324.
192. Lee HH, Yu HS, Jang JH, *et al.* Bioactivity improvement of poly(ε-caprolactone) membrane with the addition of nanofibrous bioactive glass. *Acta Biomater* 2008;4(3):622–629.
193. Chang C, Peng N, He M, *et al.* Fabrication and properties of chitin/hydroxyapatite hybrid hydrogels as scaffold nano-materials. *Carbohydrate Polym* 2013;91(1):7–13.
194. Chang MC, Ikoma T, Kikuchi M, *et al.* Preparation of a porous hydroxyapatite/collagen nanocomposite using glutaraldehyde as a crosslinkage agent. *J Mater Sci Lett* 2001;20(13):1199–1201.
195. Chen F, Wang Z-C, Lin C-J. Preparation and characterization of nano-sized hydroxyapatite particles and hydroxyapatite/chitosan nano-composite for use in biomedical materials. *Mater Lett* 2002;57(4):858–861.
196. Ishikawa Y, Komotori J, Senna M. Properties of hydroxyapatite–hyaluronic acid nano-composite sol and its interaction with natural bones and collagen fibers. *Curr Nanosci* 2006;2(3):191–196.
197. Yoshida T, Kikuchi M, Koyama Y, *et al.* Osteogenic activity of MG63 cells on bone-like hydroxyapatite/collagen nanocomposite sponges. *J Mater Sci Mater Med* 2010;21(4):1263–1272.
198. Zhang Y, Venugopal JR, El-Turki A, *et al.* Electrospun biomimetic nanocomposite nanofibers of hydroxyapatite/chitosan for bone tissue engineering. *Biomaterials* 2008;29(32):4314–4322.
199. Zou C, Weng W, Deng X, *et al.* Preparation and characterization of porous β-tricalcium phosphate/collagen composites with an integrated structure. *Biomaterials* 2005;26(26):5276–5284.
200. Matsuno T, Nakamura T, Kuremoto K, *et al.* Development of β-tricalcium phosphate/collagen sponge composite for bone regeneration. *Dent Mater J* 2006;25(1):138–144.
201. Takahashi Y, Yamamoto M, Tabata Y. Enhanced osteoinduction by controlled release of bone morphogenetic protein-2 from biodegradable sponge composed of gelatin and β-tricalcium phosphate. *Biomaterials* 2005;26(23):4856–4865.

202. Chazono M, Tanaka T, Komaki H, *et al.* Bone formation and bioresorption after implantation of injectable $\beta$-tricalcium phosphate granules–hyaluronate complex in rabbit bone defects. *J Biomed Mater Res A* 2004;70(4):542–549.
203. Lee YM, Park YJ, Lee SJ, *et al.* Tissue engineered bone formation using chitosan/tricalcium phosphate sponges. *J Periodontol* 2000; 71(3):410–417.
204. Panzavolta S, Fini M, Nicoletti A, *et al.* Porous composite scaffolds based on gelatin and partially hydrolyzed $\alpha$-tricalcium phosphate. *Acta Biomater* 2009;5(2):636–643.
205. Peter M, Binulal NS, Soumya S, *et al.* Nanocomposite scaffolds of bioactive glass ceramic nanoparticles disseminated chitosan matrix for tissue engineering applications. *Carbohydrate Polym* 2010; 79(2):284–289.
206. Kim HW, Song JH, Kim HE. Bioactive glass nanofiber–collagen nanocomposite as a novel bone regeneration matrix. *J Biomed Mater Res A* 2006;79(3):698–705.
207. Marelli B, Ghezzi CE, Barralet JE, *et al.* Three-dimensional mineralization of dense nanofibrillar collagen–bioglass hybrid scaffolds. *Biomacromolecules* 2010;11(6):1470–1479.
208. Mozafari M, Moztarzadeh F, Rabiee M, *et al.* Development of macroporous nanocomposite scaffolds of gelatin/bioactive glass prepared through layer solvent casting combined with lamination technique for bone tissue engineering. *Ceram Int* 2010;36(8): 2431–2439.
209. Couto DS, Alves NM, Mano JF. Nanostructured multilayer coatings combining chitosan with bioactive glass nanoparticles. *J Nanosci Nanotechnol* 2009;9(3):1741–1748.
210. Chen DZ, Tang CY, Chan KC, *et al.* Dynamic mechanical properties and *in vitro* bioactivity of PHBHV/HA nanocomposite. *Compos Sci Technol* 2007;67(7–8):1617–1626.
211. Noohom W, Jack KS, Martin D, *et al.* Understanding the roles of nanoparticle dispersion and polymer crystallinity in controlling the mechanical properties of HA/PHBV nanocomposites. *Biomed Mater* 2009;4(1):015003.
212. Saadat A, Behnamghader A, Karbasi S, *et al.* Comparison of acellular and cellular bioactivity of poly 3-hydroxybutyrate/hydroxyapatite nanocomposite and poly 3-hydroxybutyrate scaffolds. *Biotechnol Bioprocess Eng* 2013;18(3):587–593.

213. Misra SK, Mohn D, Brunner TJ, *et al.* Comparison of nanoscale and microscale bioactive glass on the properties of P(3HB)/bioglass composites. *Biomaterials* 2008;29(12):1750–1761.
214. Hajiali H, Karbasi S, Hosseinalipour M, *et al.* Preparation of a novel biodegradable nanocomposite scaffold based on poly(3-hydroxybutyrate)/bioglass nanoparticles for bone tissue engineering. *J Mater Sci Mater Med* 2010;21(7):2125–2132.
215. Misra SK, Nazhat SN, Valappil SP, *et al.* Fabrication and characterization of biodegradable poly(3-hydroxybutyrate) composite containing bioglass. *Biomacromolecules* 2007;8(7):2112–2119.
216. Barrett CE, Cameron RE, Best SM. Bioceramic and biopolymer nanocomposite materials for use in orthopedic applications, Chapter 2, in: Ramalingam M, Vallittu P, Ripamonti U, *et al.* editors. *Tissue Engineering and Regenerative Medicine: A Nano Approach*. Boca Raton, FL: CRC Press; 2013; pp. 19–42.
217. Tang R, Nancollas GH, Orme CA. Mechanism of dissolution of sparingly soluble electrolytes. *J Am Chem Soc* 2001;123(23):5437–5443.
218. Pan J, Han X, Niu W, *et al.* A model for biodegradation of composite materials made of polyesters and tricalcium phosphates. *Biomaterials* 2011;32(9):2248–2255.
219. Han X, Pan J. A model for simultaneous crystallisation and biodegradation of biodegradable polymers. *Biomaterials* 2009;30(3):423–430.
220. Han X, Pan J, Buchanan F, *et al.* Analysis of degradation data of poly(L-lactide–co-L,D-lactide) and poly(L-lactide) obtained at elevated and physiological temperatures using mathematical models. *Acta Biomater* 2010;6(10):3882–3889.
221. Wang Y, Pan J, Han X, *et al.* A phenomenological model for the degradation of biodegradable polymers. *Biomaterials* 2008;29(23):3393–3401.
222. Place ES, Evans ND, Stevens MM. Complexity in biomaterials for tissue engineering. *Nat Mater* 2009;8(6):457–470.
223. Havener MB, Brown LS, Darmoc MM, *et al.* Improvements in healing with a bioactive bone graft substitute in a canine metaphyseal defect. *55th Annual Meeting of the Orthopaedic Research Society*, Poster No. 508; 2009.
224. Fu S, Ni P, Wang B, *et al.* Injectable and thermo-sensitive PEG–PCL–PEG copolymer/collagen/*n*-HA hydrogel composite for guided bone regeneration. *Biomaterials* 2012;33(19):4801–4809.

225. Meretoja VV, Tirri T, Malin M, *et al.* Ectopic bone formation in and soft-tissue response to P(CL/DLLA)/bioactive glass composite scaffolds. *Clin Oral Implants Res* 2014;25(2):159–164.
226. Nie L, Chen D, Fu J, *et al.* Macroporous biphasic calcium phosphate scaffolds reinforced by poly-L-lactic acid/hydroxyapatite nanocomposite coatings for bone regeneration. *Biochem Eng J* 2015;98:29–37.
227. Sahoo NG, Pan YZ, Li L, *et al.* Nanocomposites for bone tissue regeneration. *Nanomedicine* 2013;8(4):639–653.
228. Qu D, Mosher CZ, Boushell MK, *et al.* Engineering complex orthopaedic tissues via strategic biomimicry. *Ann Biomed Eng* 2015; 43(3):697–717.
229. Pathak R, Amarpal A, Aithal HP, *et al.* Bone tissue engineering: Latest trends and future perspectives. *Adv Anim Vet Sci* 2015; 3(4S):9–22.
230. Baler K, Ball JP, Cankova Z, *et al.* Advanced nanocomposites for bone regeneration. *Biomater Sci* 2014;2:1355–1366.
231. Behari J. Piezoelectricity in bone, Chapter 2, in: Behari J, editor. *Biophysical Bone Behaviour: Principles and Applications.* Singapore: John Wiley & Sons; 2009; pp. 53–102.
232. Shahini A, Yazdimamaghani M, Walker KJ, *et al.* 3D conductive nanocomposite scaffold for bone tissue engineering. *Int J Nanomed* 2014;9:167–181.
233. Das A, Botchwey E. Evaluation of angiogenesis and osteogenesis. *Tissue Eng Part B Rev* 2011;17(6):403–414.
234. Azami M, Tavakol S, Samadikuchaksaraei A, *et al.* A porous hydroxyapatite/gelatin nanocomposite scaffold for bone tissue repair: *In vitro* and *in vivo* evaluation. *J Biomater Sci Polym Ed* 2012; 23(18):2353–2368.
235. Karageorgiou V, Kaplan D. Porosity of 3D biomaterial scaffolds and osteogenesis. *Biomaterials* 2005;26(27):5474–5491.
236. Jabbarzadeh E, Starnes T, Khan YM, *et al.* Induction of angiogenesis in tissue-engineered scaffolds designed for bone repair: A combined gene therapy–cell transplantation approach. *Proc Natl Acad Sci USA* 2008;105(32):11099–11104.
237. Buschmann J, Welti M, Hemmi S, *et al.* Three-dimensional co-cultures of osteoblasts and endothelial cells in DegraPol foam: Histological and high-field magnetic resonance imaging analyses of pre-engineered capillary networks in bone grafts. *Tissue Eng Part A* 2011;17(3–4):291–299.

238. Fedorovich NE, Haverslag RT, Dhert WJ, *et al.* The role of endothelial progenitor cells in prevascularized bone tissue engineering: Development of heterogeneous constructs. *Tissue Eng Part A* 2010;16(7):2355–2367.
239. Mercado-Pagán ÁE, Stahl AM, Shanjani Y, *et al.* Vascularization in bone tissue engineering constructs. *Ann Biomed Eng* 2015;43(3):718–729.
240. Gaharwar AK, Peppas NA, Khademhosseini A. Nanocomposite hydrogels for biomedical applications. *Biotechnol Bioeng* 2014;111(3):441–453.
241. Liu X, Zhao K, Gong T, *et al.* Delivery of growth factors using a smart porous nanocomposite scaffold to repair a mandibular bone defect. *Biomacromolecules* 2014;15(3):1019–1030.
242. Thavornyutikarn B, Chantarapanich N, Sitthiseripratip K, *et al.* Bone tissue engineering scaffolding: Computer-aided scaffolding techniques. *Progr Biomater* 2014;3(2):61–102.
243. Sherwood JK, Riley SL, Palazzolo R, *et al.* A three-dimensional osteochondral composite scaffold for articular cartilage repair. *Biomaterials* 2002;23(24):4739–4751.
244. Kalita SJ, Bose S, Hosick HL, *et al.* Development of controlled porosity polymer–ceramic composite scaffolds via fused deposition modeling. *Mater Sci Eng C* 2003;23(5):611–620.
245. Eosoly S, Brabazon D, Lohfeld S, *et al.* Selective laser sintering of hydroxyapatite/poly-ε-caprolactone scaffolds. *Acta Biomater* 2010;6(7):2511–2517.
246. Lee H, Kim Y, Kim S, *et al.* Mineralized biomimetic collagen/alginate/silica composite scaffolds fabricated by a low-temperature bio-plotting process for hard tissue regeneration: Fabrication, characterisation and *in vitro* cellular activities. *J Mater Chem B* 2014;2:5785–5798.
247. Chia HN, Wu BM. Recent advances in 3D printing of biomaterials. *J Biol Eng* 2015;9:4.
248. Hollister SJ. Porous scaffold design for tissue engineering. *Nat Mater* 2005;4(7):518–524.

# Index


## A

$\alpha'$–$\alpha$ transition, 82
$\alpha$–$\beta$ transition, 82
$\alpha'$-$Ca_3(PO_4)_2$ (super $\alpha$-TCP, $\alpha'$-TCP), 81
acrylic bone cement, 153
acrylic polymer, 151
Acrylic polymer-based cements, 152
adhesion, 114, 124–125, 127–128
adhesive bond strength, 114
agglomerated nanobioceramics, 15
alkoxysilane, 158
allografts, 2
$\alpha$-tricalcium phosphate ($\alpha$-TCP), 79–80
aminopropyltriethoxysilane, 41
amorphous calcium phosphate (ACP), 83
amorphous silica, 23, 32, 34–35
anti-bacterial activity, 174
antibiotic activity, 159
antibiotic-loaded acrylic bone cement, 159
anti-tumour action, 174
Anti-washout cement, 169
apatite cement, 162
Apatite formation, 158
apatite-forming ability, 219–220, 227, 231
apatite mineralisation, 193–194
apatite seed crystals, 169
aragonite, 173
artificial bone-filling materials, 98
artificial bones, 214, 223, 225
$\alpha$-TCP-based apatite cement, 164
$\alpha$-TCP structure, 87
$\alpha$-tricalcium phosphate ($\alpha$-TCP), 157
autograft, 2, 152

## B


barium sulphate ($BaSO_4$), 154
bending and bulge testing, 124
benzoyl peroxide (BPO), 154
beta-tricalcium phosphate ($\beta$-TCP), 79–80, 152
binary system of the phase diagram for $CaO–P_2O_5$, 82
bioactive, 3–4, 10, 13–14
bioactive glass, 37–40, 158
bioactivity, 213–214, 216, 230
biocompatibility, 30, 98, 158
biocomposite, 69
biodegradable, 3–4, 14, 248, 251
biodegradable characteristics, 98
biodegradable polymer, 241–242, 245–247, 250, 252, 256, 258–264, 268, 271, 274, 278
Bioglass®, 220–221, 223, 228
bioinert, 3–4, 13
biomaterials, 24
biomimetic coatings, 111
Bioverit®, 220, 222, 231
bond entropy, 96
bone, 24, 36–39, 239–242, 245, 250, 253–255, 265, 269–270, 272, 274, 276, 278
bone and periodontal regeneration, 188, 190
bone and periodontal tissue engineering, 198
bone-bonding, 214–215, 219, 226, 229, 231
bone cement, 151
bone cement implantation syndrome, 156
bone formation, 189
bone graft, 152
bone-like apatite, 218–220, 229, 231
bone resorption, 189
bone substitutes, 214
bone tissue engineering, 40
brushite cement, 162, 167
$\beta$-TCP crystalline structure, 90
$\beta$-TCP porous body, 84


## C


$Ca_{18}Mg_3(PO_4)_{14}$, 92
$Ca_{19}Mg_2(PO_4)_{14}$, 92
$Ca_{20}MI_2(PO_4)_{14}$ (MI:Na,K), 93
calcite, 173
Calcium, 169
calcium carbonate ($CaCO_3$), 173
calcium-deficient hydroxyapatite (cd-HAp: $Ca_9(HPO_4)(PO_4)_5OH$), 160
calcium ion, 158
calcium phosphate, 109–110, 113–114, 119, 121–123, 126–128, 132–134, 137
calcium phosphate cement, 151–152, 160
calcium phosphate nanoparticles, 242–244, 252–254, 256, 258, 260–264, 267–272, 274, 276, 278
calcium phosphates, 108, 130, 135–136
calcium sulphate, 169
carbonate, 38
Carbonate apatite cement, 172
carbonate ions, 173
carboxyl groups, 169
cell behaviours, 11
cellular uptake, 30–32
cementogenic differentiation, 195
Cerabone®A–W, 220–221, 223, 231
ceramic, 50, 56, 68

Ceravital®, 220–221, 231
Characteristics of Nanobioceramics, 6
Chelate-setting cement, 173
chelating, 169
Chemical vapour deposition, 112
chitosan, 159
citric acid, 168
coating, 50, 61–69
composite, 56, 59–61, 63, 65, 67–68
compressive strength, 86, 153–154
co-templating methods, 206
crystal growth, 169
crystalline inflammatory response, 170
crystalline structure, 87

## D

degradable polymer, 239
diametral tensile strength (DTS), 161
dicalcium phosphate anhydrous (DCPA; $CaHPO_4$), 160
dicalcium phosphate dihydrate (DCPD; $CaHPO_4 \cdot 2H_2O$), 160
dissolution–precipitation reaction, 164
distribution, 35–36
drug-delivery, 26–27

## E

elasticity, 156
electrodeposition, 63–65, 67–68
electrophoretic, 68
Electrophoretic deposition (EPD), 65–67, 111
electrospinning, 42
exothermic reaction, 156

## F

fertilisers, 80
fibronectin, 11–12
fluidity, 156
fluoroapatite, 80
fracture toughness, KIC, 154, 162

## G

gelling agent, 170
gentamicin, 159
glass-ceramic A–W, 221, 229
grain boundaries, 7–8, 12

## H

haemolysis, 34
Hall–Petch relation, 9
Heinz, Brömer, 221
Helmholtz energy, 96
high surface area, 8, 14
Höland, 230
hydroxyapatite (HAp), 37, 40, 50–69, 80–81, 108–114, 119–120, 122–123, 125–126, 128–129, 132, 136, 152

## I

IGF-I, 100
inherent properties, 7, 10
inositol phosphate (IP6), 174
interconnected pores, 170
internal energy of the crystal, 96

## K

Kokubo, 219, 221
kyphoplasty, 153

## L

large pores, 191
Larry, Hench, 214
lattice constants, 87, 90, 95
lattice constants of Mg-$\beta$-TCP, 92
lattice constants of Na-$\beta$-TCP and K-$\beta$-TCP, 94
Living systems, 11
load-bearing repairs, 162
loosening, 157

## M

macropores, 85, 170
MBG scaffolds, 195
$\gamma$-methacryloxypropyl trimethoxysilane, 158
mechanical properties, 153
mechanochemical method, 84
mesenchymal cells of rats (MC3T3-E1), 100
Mesoporous bioactive glass, 187, 189
mesoporous silica, 23
mesoporous silica nanoparticles, 25–28, 32, 34
mesoporous structure, 191, 194
metastable phases, 164
methylmethacrylate (MMA), 153
Mg-$\beta$-TCP, 91
micro-adhesion testing, 128
microindentation, 126
micropore distribution curve, 85
micropores, 85
micro-tensile adhesion testing, 114
micro-tensile testing, 124
molecular behaviour, 11
monoclinic system, 87
musculoskeletal, 50

## N

nanobioceramics, 1–2, 5–13, 15–17
nanocoatings, 106–107, 109–111, 113–115, 117, 119–128, 130–132, 134, 136–139
nanocomposite, 136–138, 239–240, 242, 244, 251–252, 254–256, 258–260, 262–267, 270–272, 274, 277–279
nanofabrication, 16
nanofeatures, 12–13
nanofibrous scaffold, 14–15
nanoindentation, 124–126
nanolaminated, 106, 136–139
nanoparticles, 27
nanorods, 23, 29
nanostructured surface, 7, 13
nanotechnology, 5–6
nanotopography, 12
nanotoxicology, 16
N,N-dimethyl-para-toluidine (DMPT), 154
nucleation, 169

## O

orthopaedic, 50, 55, 58–59, 68
osteo, 105
osteoclastic bone resorption, 173
osteoconduction, 215, 228
osteoconductive, 152
osteoconductive characteristics, 98
osteoconductivity, 154, 158
osteogenic, 152

osteogenic/cementogenic differentiation, 205
osteogenic differentiation, 195
osteointegration, 107, 109, 115, 122, 128, 132–133, 136–137
osteoporosis, 187–189
osteoporotic bone, 206
osteoporotic bone defects, 199
osteoporotic bone fractures, 189

## P

Peptide fragments, 14
periodontal/bone tissue engineering, 193
periodontal ligament cells, 190
physical vapour deposition, 111–112
pin-on-disk testing, 124
plasma spraying, 61–63, 65, 67–68, 111, 113–114
poly(lactic-co-glycolic acid) (PLGA), 159
polymerisation, 156
polymers, 248, 251
polymethyl-methacrylate (PMMA), 152
polymorphs, 81
polyurethane foam, 171
pore structure, 157
porogen, 170
porosity, 86
precipitation method, 83
promote, 59
pseudo-apatite lattices, 87
pull-out test, 124
Pulsed laser deposition, 112
push and pull-off methods, 114
PVD, 112
pyrophosphate, 169

## Q

quantum dot, 28, 30

## R

Radical polymerisation, 155
reconstruction, 151
regeneration of osteoporotic bone defects, 200
relative proliferation ratio, 100
resorbability, 173
RF magnetron sputtering, 111
rhombohedral system, 90, 95

## S

SBF, 216–220, 223, 226, 231
scaffolds, 50, 56, 58–61, 68–69, 190, 241, 250, 262–266, 268–269, 272, 274–277
scanning electron microscope (SEM), 164
scratch testing, 124
setting mechanism, 152, 163
setting properties, 161
setting reaction, 164
Si-$\alpha$-TCP, 88
Si-$\beta$-TCP, 95
signalling pathways, 5, 11
silanol group, 158
silica, 23–26, 28, 30
silicate ion, 38, 41
Silver, 159
simulated body fluid, 216
sodium alginate, 170
sodium carboxymethyl cellulose (CMC), 157
sol–gel, 111–113, 115, 117–119, 127–128, 134
Sol–gel-derived glass, 39

sol–gel method, 84
solid-phase reaction method, 81
Solid Solution Substitution, 87
solubility, 96
$Sr^{2+}$ ions, 203
Sr ion, 194
Sr-MBG, 187, 190, 199, 205–206
Sr-MBG scaffolds, 191, 193, 198, 203
Sr or oestrogen, 200
Strontium, 187
strontium ranelate, 188
supersaturation, 164
surface roughness, 8, 12
surface wettability, 8–9
synthesizing, 54

## T

Tadashi, Kokubo, 216
Tensile strength, 154
tetracalcium phosphate (TTCP: $Ca_4(PO_4)_2O$), 160
TGF-$\beta$, 100
therapeutic Sr ions, 205
Thermal spraying, 111
thermodynamically stable phase, 163
thermodynamic behaviour, 96
thin-film, 106, 109, 113, 124
third generation biomaterials, 1, 198
tissue engineering, 40
titanium oxide, 158
tooth, 36–37
toxicity, 32, 36
transmission electron microscope (TEM), 164
transparent property, 7
tricalcium phosphate, 79, 111
triethoxysilane, 25
TTCP-DCPA-based cement, 165
two-component solid-phase reaction of $Ca_3(PO_4)_2$–$CaNaPO_4$, 93

## V

vacancy, 93
vancomycin, 159
vaterite, 41, 43, 173
vitronectin, 11–12

## W

wear testing, 124
well-ordered mesopores, 192
whitlockite, 81, 95
Wolfram, Höland, 222

## Y

Young's modulus, 60, 154

## Z

Zn-$\beta$-TCP, 100–101